The Universe of Experience

The Universe of Experience

A Worldview Beyond Science and Religion

Lancelot Law Whyte

With a new introduction by
Gary David and **Brian Rothery**

Routledge
Taylor & Francis Group
LONDON AND NEW YORK

Originally published in 1974 by Harper Torchbooks

Published 2003 by Transaction Publishers

Published 2017 by Routledge
2 Park Square, Milton Park, Abingdon, Oxon OX14 4RN
711 Third Avenue, New York, NY 10017, USA

First issued in hardback 2017

Routledge is an imprint of the Taylor & Francis Group, an informa business

Library of Congress Catalog Number: 2003050775

Library of Congress Cataloging-in-Publication Data

Whyte, Lancelot Law, 1896-1972.
The universe of experience : a worldview beyond science and religion / Lancelot Law Whyte.
p. cm.
Originally published: New York : Harper & Row, 1974.
Includes bibliographical references and index.
ISBN 0-7658-0505-7 (alk. paper)
1. Cosmology. 2. Science—Philosophy. 3. Human beings. 4. Experience. I. Title.

BD511.W53 2003
113—dc21

2003050775

ISBN 13: 978-1-138-53925-9 (hbk)
ISBN 13: 978-0-7658-0505-8 (pbk)

To Any Reader

On whom his work makes an impression, I quote Dostoevski's remark after his Pushkin Memorial Address (1880), which made a profound impact on his audience: "It did not make an impression by its merits, nor by any talent in its exposition . . . but by its sincerity, and I will even say by some irresistible power in the facts displayed, notwithstanding its brevity and incompleteness."

Contents

Introduction to the Transaction Edition

In the fall of 2001, Brian Rothery journeyed from Ireland to Boston to look at the private papers of L.L.Whyte, stored at Boston University. Brian was the first scholar to research the Whyte papers. He was faced with forty boxes of documents and some audiotapes. In box number 36 was a package labeled "Private Diaries." Inside was a large white envelope, sealed with wax and with a warning that it was not to be opened until after May 1983. Before that date had been reached, Whyte's widow Eve extended the restriction by a further five years until 1988, the year in which she was to die. Inside the envelope were several student-type notebooks, covering the period from 1948 to his death in 1972, almost twenty-five years. Addressed to future biographers and researchers, the diaries revealed that Whyte was attempting to express a unified theory more comprehensive than Einstein's Theory of Relativity, which included far-reaching meanings for the human psyche, and encompassed psychology, philosophy, and physics. They further revealed that this private undertaking seemed to be Whyte's main passion, from which writing books was a mere "respite." The unified theory included a mathematical expression of a pervasive order-generating tendency at work at all levels of matter, life, and mind. Theodore Roszak, the well-known author, and Whyte's friend, visited him for the last time in a London hospital after his 1972 heart attack. Thirty years later Brian interviewed Roszak: "I recall Lance telling me that he believed his theories of form in nature required a mathematical formulation—something as simple as Newton's laws of gravity. He never gave up hope that a mathematics of 'global variables' would yet be invented to give his morphic vision a rigorous, experimental expression. He said he had struggled to find that formulation to the point of profound depression that bordered on suicide." Theodore added, "I remember being struck by this—it made me wonder about the domineering role of mathematics in science."

Who was this man who lived an extraordinary life by any standard? He met and explored his concerns with some of the most famous scientists, philosophers, and personalities in the world. Bertrand Russell wrote him a letter that read in part, "When I read anything of yours, I find myself thinking that what you say is likely to be both true and important." For someone with such connections and thirteen books published in his lifetime, he was virtually ignored in his native country and had a small readership anywhere, except in the United States for *The Next Development in Man.* He had only short periods of what most of us would regard as normal work, the London bank, his part in the Whittle jet engine project, the wartime Ministry of Supply. No long-term employment, no steady journalism or regular book royalties. His main occupation was to foster *unity* wherever he could.

Out of Whyte's private papers, a picture of the man unfolds to show the affective fires of meaning that fueled his *oeuvre.* He emerges as a passionate, knowledgeable, vital, ill-at-ease, and sometimes anguished observer-participant in the ongoing drama of human life on earth. In his private papers he wrote, I have found life fascinating, rich and rewarding. I have loved and been loved. I see much beauty. But I am not surprised that some consider human life tragic or 'petty.' They may have been born with less vitality or been less fortunate than I. It is they that are poor, not life. Come what may, personal disasters or a painful end, my experience has been rich. I have found life at many moments as beautiful as my nature could bear."

He was a master at painting the big picture. Like an empathic alien, he seemed to be trying to report to anyone who would listen what he sees from outside the gravitational pull of the epistemic systems of the past. In a paper given in 1969, he said,

> This half-century has shown that only some unprecedented factor, realizing human potentialities in a new manner, can arrest the progressive collapse of civilization and the final corruption of those values, which are still in some degree effective. It is possible, I suggest, that tomorrow holds a major surprise of this kind. Anyhow I am certain that no one has the right to assert that he knows that this view is mistaken . . . This is not an intellectualist or idealist exercise, but a personal, historically oriented, manifesto, one mark-ing a hundred years advance: coordinative, not commu-nist. I am not considering the reformulation of ideals in words, or hierarchical systems of verbally expressed val-ues, but something more fundamental: the rediscovery and rejoicing in the daily living into action of vital values *as the result of a changed human condition* or state of organization, individual and social, the development of which all will facilitate who are not perverted. The "new synthesis" to which I point is not only a philosophical and

> scientific unification of rational knowledge. It is that and much more: a human synthesis, an organic co-adaptation, reorganization, and harmonization of human capacities in some respects equivalent to, but more profound than, the spread of a new universal and unifying "religion," in the traditional sense of a doctrine involving transcendental features and lacking coherence with scientific knowledge.[1]

Brian's sense of significance grew as he came upon a document —half-typed, half-written—titled, "An Anonymous Testament." Reading it like an archeologist who comes upon an unsigned artifact from another time, he saw, "This is a confession of faith. I make it anonymously. I doubt I will ever say or note anything more basic. Aware of myself as believing I can make some contribution to others. I do not claim ability to live my faith. Moralizing for others is hypocritical and empty. Only silent example unaware of itself, is effective. I am being as sincere as I can be."

Lancelot Law Whyte (known as Lance by friends and family) actually did live by his faith. His silent example was that of an original thinker unafraid to buck the tides and trends of his time. Yet, his almost ruthless self-reflexive awareness compelled a rethinking and reworking of his life-long themes, personal, scientific, human; revising his life and his writings in light of his latest insights and perceptions; coining new terms and giving new meanings to old ones. He gave discriminating welcome to phenomena like the American youth movement of the 1960s and showed interest in human mental abilities that were unexplained by common sense and accepted "laws of nature." All of these were part of his silent example of dedication and fidelity to unity, creativity and to his own world-view. He was a model of what a scientist-philosopher could embody in the twenty-first century—creative, non-nihilistic, technically knowledgeable and informationally up-to-date.

The Universe of Experience, is, in his own words, ". . . a paean to unity, sung by a skeptic." It was not written with just scientists in mind, "It attempts to share with the nonspecialist readers the author's view of a radical metamorphosis of the psyche already under way, particularly in the West." In it, one can also experience a consciously self-reflexive mind in action, in relation to all his previously held convictions—validating some, and revising others. This book was a manifesto that unified his thirteen previous books including such important works as *The Next Development in Mankind* (1944, 2003), *The Unitary Principle in Physics and Biology* (1949), *Accent on Form* (1954), *The Unconscious Before Freud* (1960), *Focus and Diver-*

sions (1963), and *Internal Factors in Evolution* (1965). It was his farewell public document that laid out, with a sense of great urgency, some of the key factors that must be addressed if we are to survive and thrive as a species. He said, "I do not suggest that (as) a *probable* outcome. In our ignorance regarding historical processes that term has no useful meaning; who can know what is probable tomorrow, since history is a web of surprises?" So his forecasts may be regarded as *possibilities* and not predictive of probable outcomes.

What are some of them, and what can we say now more than thirty years later?

> 1. [I]f total disaster can be held off, there is a possibility of a worldwide consensus emerging soon, and spreading rapidly, concerning what it should mean to be a human being in the late twentieth century, and, on steps to implement this consensus, . . . an unconscious tradition has since 1914 been preparing the ground for a sudden emergence into general awareness in many countries that such a consensus is both necessary and possible. Moreover I believe that signs of this are already visible.

While we may doubt such a consensus in any political sense, there are gleamings in the grass of an informal world community of those who have glimpsed the significance of unity in its scientific, aesthetic, philosophical, religious, and "mystical" aspects. They are those for whom wholeness, or oneness is a "fact" that we accept; unity is a condition that we can create and recreate whenever circumstances provide the opportunity. In the scientific community, for example, there is an emerging awareness of what Edward O. Wilson called "consilience" as a key to unification among the various epistemic formations that have been separate departments up to now. Consilience literally means "jumping together." It was coined by William Whewell in 1840. Wilson wrote, "literally a 'jumping together' of knowledge by the linking of facts and fact-based theory across disciplines to create a common groundwork of explanation."[2]

There are attempts to unify perpetrators and victims of crimes by bringing them together into an atmosphere of trust and revelation, rather than one of adversarial "justice." In South Africa there is the South African Truth and Reconciliation Commission which was set up by the Government of National Unity to help deal with what happened under apartheid. In Rwanda there is the Ministry of Reconciliation called the Umuvumu Project to help victims of crimes of genocide. There are conferences and ongoing groups dedicated to "Restorative Justice" in different parts of the world, in all walks of community life. Physicist David Bohm brought forth a process called "dialogue" in which groups of people could suspend their deepest values and talk together. These are a few examples, but all such formations are unity-seeking and value-based that take into account the affective dimension of human experience—the dimension that Whyte pointed to as having been devalued during the period of the 'dissociation' of European man (see his account in *The Next Development in Mankind*).

> 2. The structure of the universe at all levels will be known to be less arbitrary than had previously been imagined.

We are beginning to see great advances in biology such as the genome project, as well as confirmations in cosmology. The *New York Times* headlined, "New Map Unveiled of Universe at Start of Time." It was a declared confirmation of the Big Bang Theory in greater detail than ever before in which the basic parameters that characterize the universe, including its age, geometry, composition and weight, have entered the realm of the known.

> 3. I consider that it would be important if, during a decade or so, those interests which benefit from the promotion of violence could be identified, publicly exposed, and brought to trial by a World Court empowered for this task.

We seem to be far from such a World Court at this point, but there are those hard at work exposing the roots of violence in our communities, including schools, such as the program in understanding and managing shame and humiliation developed by psychiatrist Donald Nathanson of the Silvan Tomkins Institute of Philadelphia. And there are whole schools dedicated to studying peace and conflict in international relations in an attempt to avoid violence. The practices of corporate culture are coming under scrutiny as never before. The press, at its best, also provides a kind of "world court." We're speaking of true journalists and reporters. In one of the notes in the back of the book (p. 147) Whyte notes, "I have often noted that a very few writers, say five to ten, have prepared the way for a necessary and ultimately decisive switch of public opinion in England or America. These are the salt of the earth, but they are usually given their opportunity by editors or owners of the media. At the present moment the existence in the Western world of some hundred responsible men and women of good will and good judgment—if they exist in positions of influence today—facilitating the spread of a world consensus by selecting appropriate reporters and commentators could be of decisive influence. . . a few hundred newswriters and commentators now have in their hands human fate in this century. Great themes must be brought to birth by a few."

4. The decades from 1950 onwards will be recognized as marked by a change in evolutionary philosophy: the gradual discarding of an unduly narrow view of the mechanism of the evolution of species as due only to the external or Darwinian adaptive selection of matured forms resulting from haphazard mutations.

This is being born out everyday in the current climate of evolutionary science where evolution is seen as much more than merely adaptive. Darwin himself wrote, "I am convinced that Natural Selection has been the most important, but not the exclusive, means of modification." And one no less than Stephen Jay Gould in a *Scientific American* article wrote, "Yet powerful though the principle may be, natural selection is not the only cause of evolutionary change (and may, in many cases, be overshadowed by other forces)." Whyte's construct of a 'vital surplus' can be very valuable in today's evolutionary theories. (See the definition of vital surplus below.)

5. I consider that the possibility of a unification of a physiological theory of brain processes and a psychological theory of emotional and cognitive processes is implied by the world view.

Unknown to Whyte at the time, the profound work of psychological researcher-philosopher Silvan S. Tomkins (1911-1991) was in process of development. Tomkins, through a Herculean personal effort, starting in the mid-1950s, laid the ground work of affect psychology which brought together the elements that Whyte mentioned. His four volumes titled, *Affect, Imagery and Consciousnes*" stretched in publication from 1962 to 1992. His affect and script theory fulfilled Whyte's possibility of psychological unification of affect and cognition, body and mind. We have seen an exponential growth in the brain sciences, and a plethora of new books on the findings offer grounds for new insights into the structural fabric of human "minding."

6. The period ahead, in my view, will be marked by its emphasis on *hierarchies of morphic processes* in the inorganic and the organic realms, on *man as a coordinated organism*, when not sick *perpetually creating new unities*, new unions of contrasts, and, above all, endowed with a capacity for *joy*.

The hierarchical view has come into great favor in biology, astronomy, in the life sciences, and in the social sciences. In biology, hierarchical structures are described as assemblies of molecular units or their aggregates embedded within other particles or aggregates that may, in turn, be part of even larger units of increasing levels of organization. In regard to joy, Whyte states as article 1 of a new universal declaration of rights that we each possess the faculty of experiencing perfection and must not be frustrated in enjoying it. Silvan Tomkins, too, postulated that a central image or goal of human mental life (when not pathological) was to maximize or optimize positive affect (enjoyment-joy, and interest-excitement) and to minimize the negative affects such as (on a continuum of intensity) shame-humiliation, fear-terror, anger-rage, distress-anguish, disgust and dissmell. Joy is more than what we usually mean by "happiness." Whyte saw it as the bliss of enjoying and serving a pervasive unity. It is vitality without the discord that comes from an over-abundance of dissociated self-consciousness.

> 7. The great task of the coming decade for physics and biophysics, and perhaps for embryology and for understanding the development of the thinking brain and of language, is to discover the hierarchy of morphic rules relating all levels in all relevant processes.

The new science which this achievement would inaugurate would significantly differ from the traditional special sciences in its holistic character. It would be global and relational, that is, treating of order and disorder, patterns of relations within a totality rather than substance-bound or atomistic. We are seeing attempts to bring into consideration more general features of the psyche, or of cerebral processes conditioned by the fact that they are within the organism, into the bigger picture. We are catching glimpses of a single continuity running through all realms of the universe: the "physical," "organic," "mental," and "spiritual" (*geist*).

These are just a few of the possibilities Whyte saw. Some missed the mark in terms of specifics. For example, on he wrote, "This implies that the ordering tendency, here called *morphic* (see chapter 3) will before the end of this century be taught to children in many lands as one of the axioms of a new world view. This is my prediction, made in 1972 and placed here on record." Some were amazingly accurate, and most remain relevant and open questions today. On page 101, Whyte clearly states, "'Very well,' you may say. 'This may be right. So what? Where do we go from here?' This question is one I do not need to put or to answer. I am not here drafting a scenario for (humankind). My aim is the expression of a conviction which I believe has power and timeliness. I repeat: These words are not written for a purpose, in order that tomorrow a new social movement may be born visible to all, or a new social advance be achieved the day after. I am concerned only with the present movement of the human psyche, and I write for the countless thousands of others who feel and think as I do and will be strengthened by my words, as I am by my knowledge that those others exist."

Whyte's fellow travelers have a fertile field to harvest. In each and every one of us, as we engage in conversation, read a book, watch an artistic performance, there lies, beneath our awareness and sometimes dimly sensed between the spoken or printed words, patterns of meanings awaiting their time to emerge into conscious clarity. It is here that the movement towards unity and coordination takes place, while in the conscious there can be discord and division, like being in underwater calm while a storm wages on the surface of the sea.

If any of us is achieving a better sharing of ideas and feelings with others through such media as the Internet, then Whyte's forecasts are born out. He said that we often yearn for something more than personal which we cannot identify. I feel that I need to go somewhere, talk with someone about something, but I know not what. This can be seen as a signal from an "unconscious" seeking unification. A mother smiles, a teacher's praise, pass valuable information beyond the channels of consciousness. Here is a primary and supreme value, an acceptance of joy in what is good, communicated without words on an affective carrier wave. But now we can also send an email.

One can come away from reading *The Universe of Experience* with a cohesive and comprehensive basis for trusting our deeper selves—the selves that encompass the very ordering forces that gave rise to the universe we inhabit. No book can do it for us, but a man's life was dedicated to demonstrating what can happen if we make the leap and abandon the methods of self-protection that we developed to deal with the traumas of the childhood of humanity.

Gary David and Brian Rothery

Notes

1. Whyte, L.L. "Recover Values in a New Synthesis", Main Currents.
2. Wilson, E.O., "Consilience: the unity of knowledge", Alfred A. Knopf, NY, 1998.

Key Terms

There are key terms in *The Universe of Experience*. If those terms become familiar and understood, Whyte's vision will roll at your feet like a friendly dog. Here are a few:

Morphic—Tendency toward organic coordination - generating order, form or symmetry in ordinary 3D space. Well-formed terminal states can arise from less-formed initial ones. Everything we perceive has form and was formed sometime from something less formed. The morphic tendency leaves a specific memory (a record) of particular events as well as the process by which they were formed; the entropic does not leave such a trace.

Perconscious— All those mental processes in which two or more levels are at work; one conscious and one less so. The perconscious is the omnipresent interplay of several levels of the mind expressing many levels at once. Only the perconscious in individuals can shape a complex unified pattern.

Joy —Vitality without discord. An unsought state of grace in which life is enhanced and the imagination surpassed; the unexpected evocation of one's fullest aesthetic response; the surprising experience of perfection.

Unity—A known or unknown principle of ordering. One general process giving rise to differentiation and diversity.

Vital Surplus—Since in general the morphic tendency is cumulative and generates stable structures it produces a 'vital surplus', which is not merely adaptive and life-preserving, but life-enhancing, formative and sometimes creative. It is part of man's inborn nature to struggle to overcome whatever limits his actions and his coordination and sometimes creative. Humans suffer from surplus capacities beyond the needs of biological survival and beyond anything that could be ascribed to natural selection of favorable characteristics. Homo sapiens appears to be cursed and blessed with a restlessness springing from still unrealized potentialities, far in excess of the degree of vitality that would be biologically appropriate or adaptively most advantageous It shapes new forms of behavior which may or may not prove advantageous, and it does this *before* the evolutionary process tests them out.

Imagination—The primary capacity of the human mind of which all other mental capacities should be understood as special cases or applications. The imagination depends on the interplay of at least two levels in the human mind (perconscious). It is the faculty for forming new unities. The imagination can be defined and understood without the intellect; but any understanding of the intellect must assume some ordering generated by the imagination.

Understanding—'Thinking' means more than analytical intellectual reasoning. To experience and to act in an organically appropriate manner, man must be in possession of complete aesthetic, intuitive, philosophical, and scientific understanding of the basic features of this universe, such as matter, life, and mind. That includes, of course, understanding of his own potentialities, including the general and abstract affective energies that motivate all meanings and behaviors as well as the cognitive functions that transform those energies into action and direction. Such coordinated understanding based on unified knowledge is the criterion of biological maturity and viability for *Homo sapiens*.

Acknowledgements

After Lancelot Law Whyte died in 1972, his widow Eve tidied up the manuscript of *The Universe of Experience,* wrote a foreword, and added the following note:

Note Number 8 to Chapter 7 (See Notes chapter): L. L. Whyte had intended to discuss in more detail elsewhere "the now indispensable, potentially universal consensus of heart, mind, and will which all nonpathological individuals must necessarily accept, that alone can lead mankind safely through the next few decades."

Eve then tried to get a publisher, but found little interest. She sought the help of the young, already well-known, writer, Theodore Roszak, who had become their friend, and, through his intercession, *The Universe of Experience* was published in 1974. Later, Roszak described the poor performance of the publisher in its selling of the book as "scandalous." Shortly after publication the value of "returns" so exceeded that of the small initial advance to Eve that she owed the publisher money. If, however, Eve and Roszak had not managed to get that book published, causing a worn copy to come into our hands nearly thirty years later, it is doubtful if Lancelot Law Whyte would be republished today, as the impact of reading it and *The Unconscious Before Freud* on top of the already long-felt impact of *The Next Development in Man*, hardened our resolve to rescue and republish his books.

Two important U.S. friends and admirers of Lance were Boston University's curator of the university's Special Collections Dr. Howard Gotlieb, and Dr. John Silber, who has since become the chancellor of the university. After Lance's death, Howard Gotlieb visited Eve and secured all the manuscripts and private papers and literary rights. This was very important as it smoothed our path to republishing, so that when we approached Boston University we found ourselves pushing on the open doors of both Howard Gotlieb and John Silber.

We were struggling with the task of scanning in old copies and finding designers and printers within our modest budget, when another of Lance's admirers came to our rescue. Ninety-year-old Richard M. Gummere, Jr. (Buzz) found our Whyte web pages and picked up the phone and called yet another Whyte admirer, his friend Irving Louis Horowitz, himself a best-selling author, who was chairman of Transaction Publishers. Transaction took over the considerable task of publishing.

Of great help to us on family history and related matters was Lance's niece, Anne Whyte (married as Anne Thoms), who is a retired journalist living in Edinburgh, not far from the original Whyte family home. Also helping us with family matters, were Eve's relatives, the Korner family, one of who, upon hearing of the plans to republish Lance, remarked,

"If only Eve were alive to hear this today."

Foreword

Lancelot Law Whyte (1896 1972) devoted a great part of his life to "unitary thought." The idea of a pervasive unity, a tendency towards order came to him in 1925, in his late twenties, and thereafter he endeavored to find a principle that could account for the existence of form and order in the inorganic world, in organic nature, and in man's mind. He has recorded how in 1929 there flashed upon him the most general concept, what he called the "Unitary principle, by which *asymmetry decreases and gives place to symmetry.*" His books and articles are, in essence, variations on this theme; they are applications and developments of the basic idea he conceived between 1925 and 1929. "My aim," he says in his account of himself, "was to seek to bring into closer relation the scientific conception of the forms of external nature and my personal sense of the forms of experience." †

Among his contemporaries, Whyte was unusual because he knew, in the full professional sense, what modern physics and biology were about, but at the same time had the aesthetic approach to experience of an artist or poet.

In the present book, his last, he brought together many of the themes that had occupied him during his life, and he sought once again to justify, from a new angle, his conviction that human emotion and thought and the nature of the universe are governed by the same law. A note left with the manuscript, dated February 6, 1972, shows that he planned a book "relatively short (30/50,000), very clear, for all scientists (and others) ...," and another note, dated midnight, Wednesday, February 23, 1972, reads: "My role of conscience in writing this book: to write as honestly as I know how, as if it were *my last word,* my testament. To be recklessly honest—To consider no reader, friend or foe .... So bequeath my best."

---

*L. L. Whyte, Focus and Diversions (New York: George Braziller, 1963), pp.170 71.
† Ibid.

Lancelot Law Whyte, my husband, was then in his seventy-sixth year. Ten years earlier he had thought, on grounds of heredity and health, that he might live to be eighty-five,* and he had planned further work in fundamental physics and a number of books concerned with one major theme, namely: "The advance towards a comprehensive unification and simplification of scientific knowledge . . . so that everyone can, in some degree, feel its truth in their own experience. Scientific knowledge must become human understanding. That is, I believe, the only way in which humanity can acquire the moral strength to control technology.

But in 1972 he was beginning to feel that his stock of days was running low; as he says in the book: "This work may be my swan song on universal themes." He worked on it during the spring and summer of 1972, and it was complete when he died suddenly on September 14.

Eve Whyte

* Ibid., p. 230.

† A brief biographical note, 1969, for use in the U.S.A.

List of Abbreviations

The following abbreviations will be used throughout the text and notes:

NDM: *The Next Development in Man* by the author. Published by Cresset in 1944.

3D: The three dimensional realm of the immediately observed (symmetrical, spatial) relation of *distance*.

S: The immediately experienced (asymmetrical, temporal) relation of *succession*.

C.C.: Coordinative conditions of cellular organisms.

1

Who has Written This and Why

An interpreter of science and religion who offers a world view overcoming nihilism

This book is a paean to unity, sung by a skeptic. He knows that beauty can deceive and that the true unity can be other than one's most superb vision of it. The greatest have made mistakes, and the more personal the enthusiasm that an idea evokes in a person who believes he is its creator, the more likely is it that the idea is corrupted by mistaken unconscious assumptions. This pervasive irony haunts all human thought, including these pages. They contain unassailable convictions and frail personal judgments. Only the universal is free from blemish, and its carrier must protect himself from an overestimation of his role by an irony which gaily soars over all dogmatisms.

But this awareness does not lessen either the earnestness of these pages or the author's daring. He has no choice but to discard false humility and to declare his purpose, which is not modest. Unlike his previous works this book offers a doctrine in simple language for everyone who cares, but above all for young-in-spirit America. Most technical terms are banished to the notes, and the powerful theme can carry nonspecialists through such technicalities as remain.

We are sick today for lack of simple ideas which can help us to be what we want to be. The basic challenge to mankind is not population, poverty, war, technology, pollution, religious or racial intolerance, or blind nationalism, but an underlying *nihilism* promoting violence and frustrating sane policies on these issues. In 1870 Nietzsche knew that nihilism stood at the door; now it is with us. Despair for mankind is nearly universal. What hope is there? The current protest of youth is basically a rebellion against a nihilistic society, but what comes next?

The author's view is that the only hope lies in the emergence of a potentially worldwide consensus of heart, mind, and will, appealing to all sane men and women everywhere, because it looks deeper into the nature of man than Marxism or any other past doctrine. The time has come for the West to speak to the world in universal terms. In chapter 7 we shall see that the emergence of such a consensus is not excluded by the horrors of the past or of the present, and the consensus, if it comes, is likely to surprise by its suddenness, timeliness, and universality. But it may take decades to mature and to be sufficiently, implemented. The world view presented here is one man's contribution to such a new uniting attitude blending emotion, thought, and action. It covers in broad outline the primary features of the universe of human experience, within and without, and is offered to all educated men and women. Important things can be said simply. But brevity does not imply dogmatism, for the irony which protects these pages leaves the way open to complementary or deeper truths which may be disclosed by others. Many allies are necessary in the development of the consensus.

But what title has the author to attempt such a task? What empowers him to seek to bring clarity into a time of confusion?

Well, for a start, he is a scientific frontiersman[1] who has spent fifty years, not tied to a chair, but using his free time to canter around the no man's land of science along the frontiers where physics, biology, and psychology face the unknown, with an eye also on philosophy.[2] This book describes the response evoked in him, as a person and as a thinker, by what he has seen as he watched raids into the vast fields of our ignorance. Yet that description may mislead. For he has been a prejudiced observer, attentive only to those deep raids which he could interpret as leading toward the unity of the sciences and of knowledge, regarded as a necessary component in a unified human being. He is a *metaphysicist* in search of unity. By all means let there be the highest objectivity possible in each realm, but, life being brief, spend it watching and perhaps aiding the great convergence toward the unity of knowledge which is to him already unmistakable. That is the author's retrospective rationalization of his intellectual life.

So he is an interpreter using his experiences in watching philosophy and science to identify the convergent paths in all realms that the intellect must take *if a unified view is to be achieved.* He attempted this long ago in *The Next Development in Man,* using a historical approach, and also elsewhere.[3] Here he simplifies the task by considering only those issues which he regards as basic for us all, here and now. The years have brought him assurance, and he can already see the point of convergence ahead. He believes that the time has arrived when a new clarity which all can share is necessary and possible. Unity, clarity, and immediacy are his guardian principles.

This essay *touches bottom* for twentieth century man. If its theme is timely it is one of many signals marking the end of "Antiman," with his hopeless relativism, and announcing "Unitary Man,"[4] which stands for a new type of man, able to be more harmonious because he has become aware of the ordering processes at all levels in nature, without and within. In this new vision man finds an absolute on which his thought can rest, for here at last subject and object are potentially fused in a single insight. History is a sequence of surprises, and the authentic advances of the human mind, bringing new and fertile clarities, are not at an end. This book suggests that the human psyche—a term here used for the potentially unified heart, mind, and will—is about to turn a corner and enjoy vistas never seen before. It attempts to share with nonspecialist readers the author's view of a radical metamorphosis of the psyche already under way, particularly in the West. This is an essay in unity. There is nothing said here which has not its aesthetic, religious, moral, philosophical, and scientific aspects, for this work traces everything to their common source: the human imagination and its organic roots.

But if he goes beyond established science, whence the writer's authority to interpret the present so as to foresee a possible future? What an ambition! Amid today's confusion no man would attempt it if he experienced any hesitation or freedom of choice, in fact unless something was compelling him to think and write as he did. Such outrageous presumption is tolerable only if the message has the ring of an authenticity born of necessity, experienced by him and clear to others. Whence this daring in our time of confusion?

The author's answer is simple but unusual, and before he makes his dread confession—which may lose him some readers' sympathy—the ground must be prepared. This book is mainly philosophical and *pre*scientific, and the next pages may seem to be a diversion, but they are not. To understand the ideas he presents one must also identify his impulse and its source, and to do that we must first take a look at the source of transcendental religion, which here means

Christianity. It is the miserable failure of Christianity which brings him to write as he does with passion and compassion, and, at a deeper level, dispassion. May the indignation now displayed give the color of life to these pages. They are written by a man who has never before expressed his anger at the pretensions and dishonesty of transcendental religion. Anger with Christianity? No longer. Anger, now outgrown, with himself at being for so long so naïve.

Every Sunday throughout the West countless parsons and priests open their sacred book and, claiming transcendental inspiration, promise their flock the blessing of God provided they truly seek Him. As individuals they have the author's tolerant understanding; we are most of us wandering in the dark. It is the institution that shocks him. The sins of Christianity have howled through the centuries, and in his own lifetime they have been such as to rob the Church of his last lingering respect. For him, as for Nietzsche, the Western God died in effect a hundred years ago, and since then the wrigglings of theologians trying to explain what they still believe have been as contemptible as the behavior of those secret communists who never came clean. Do they or do they not still believe in a transcendental divinity? The community is waiting for their answer.

The hard fact, surely as clear as daylight to all who seek the truth, is that we know nothing about the Christian transcendental God except His total indifference both to individual human suffering and to the collapse of the pseudo Christian civilization which the Church supported. The pretension to a transcendental authority has survived until recently because it was a collective hypnotic given to children and remaining with them as adults. During one period in history it served a purpose, being helpful both to believers and to their rulers. It met the desperate human need for beauty and comfort amid the strain and pain of daily life, and it brought some stability. Moreover, its early rationalist opponents, though in many respects they led the conscience of mankind, were strangely naive, for they imagined that a half-developed faculty called "reason" should, and could, lead the species. But reason could never be equal to that task, for it is not a prime mover.

Today few are so credulous as to accept a transcendental personality. Our situation is different. Faced by the dire nihilism of our time, inevitable after the bogus idealism and religion which preceded it, we need a greater honesty than popes, archbishops, and their subordinates can offer: A new human awareness is already widespread, and this will lead to novel insights, going deeper into the character of this universe and of man than any which he has ever had before.

The Western search for a unifying truth did not come to an end with Christianity, any more than with the physical theories of forty years ago. I believe that the human mind can still make basic advances. But to do so sectarian conceits such as the Christian pretension to transcendental inspiration must be explicitly cast aside. It is already seen to be hollow, but it must now be openly dropped, so that we can breathe clean air unpolluted by pretense. Some of us may think of ourselves as agnostics or atheists. But insofar as we have been born and bred in the atmosphere of Christianity in ruins, the Christian fog of self deception still does its damage: we either deceive ourselves by pretending to believe or overreact into a contempt of all religion. So, away with the fog!

Is anyone surprised at the impotence of Christianity to help mankind today? God doesn't give a damn. Away with the sentimental nostalgia for the faded beauties of the past which merely distort our vision, not only the real limitations of man but also of his real strength, *the "divine" in man:* creative bliss, the experience of perfection, the surprising joys of love all human, not divine. It has long been understood that whoever denies God asserts his own divinity. In dropping God, man recovers himself. It is time that God was put in his place, that is, *in man,* and no nonsense about it. But, to prevent misunderstanding, instead of speaking of the "divine" in man I will call it the human sense of perfection or unity. We are all restless creatures until we have adequately exercised our own capacity for perfection. Need I add that we can retain the Sermon on the Mount, Saint Paul's poem to charity, and much else, though we discard the Christian God?

Without any conscious preoccupation with this issue, for the mock battle of transcendental divinity with aggressive atheism never interested him, the author has allowed this question to work itself out unconsciously within himself throughout his life. He has some understanding of the power and appeal of transcendental religion, yet he has known that for several centuries freethinkers have led mankind. Only recently, with the bliss that accompanies an authentic discovery, new to him though perhaps long understood by others, possibly Kant and certainly Nietzsche, there emerged into his mind a clarity that will remain with him to the end: the conception of transcendental divinity is damaging to man. *The idea of God lessens man*. I assert this in the name of beauty, truth, goodness, and joy, before the bar of history. As Nietzsche said in his *Joyful Wisdom:* ". . . perhaps man will rise higher when he stops projecting himself into a God." Today man needs to understand and to accept the deepest level of his mind, the impersonal potentially all powerful factor that links him with all men, the source of joy. In this clarification lies the transforming consummation of religion, not its reduction.

Whence came this unnatural idea of a Transcendental Agent? The ability of the human mind to make fundamental advances—a miracle in itself—is shown in the fact that whereas in 1872 only a few pioneers knew the answer, in 1972 it is common knowledge, though some may still prefer to deceive themselves. Belief in a transcendental divinity arose from a misinterpretation of intimations from the less conscious levels of the mind. (The terms "conscious" and "unconscious" are misleading; this will be considered later.) God is in the unconscious, *is* the unconscious, perhaps. As Jung said, "We are unable to distinguish whether God and the unconscious are two different entities," and Hinduism, recognizing the impersonality of the Divine Being, identified it with the deeper Self of man.

But the deepest level of the healthy unconscious mind is much more than God ever was, as we shall see in specific detail which cannot be gainsaid. Today most know that the assumed transcendental origin of the most powerful unifying intimations is an unfortunate misinterpretation, for it is not only redundant and misleading, but it lessens man, taking the best away from him and making it less accessible, except by the gracious intervention of a hypocritical priesthood. At least it has been so in the orthodox Christian churches; the Quakers and Unitarians perhaps knew better.

Here a caution may be in place. The existence of the unconscious mind and the use of this conception do not imply a retreat from clarity or a neglect of intellectual discipline. Some have contrasted a blind and helpless unconscious with the precision and elegance of the intellect. This is a misinterpretation of one of the greatest discoveries ever made by man, an old one clinched in this century: that the conscious mind does not know all that the total mind knows. The unconscious mind is not blind, and its several levels sustain and nourish the intellect and the imagination. There exist interacting, cooperative levels of the healthy mind, still to be understood, toward which our first tentative approximations, "conscious" and "unconscious," only crudely point. But back to our theme.

The obvious must be restated, for it is so important. The vaunted transcendental inspiration of Christianity was a flow of intimations from less conscious processes in the minds of Jesus and of his more inspired followers, interpreted in terms of the Messianic tradition of their community. The source of Mohammed's inspiration was the same, and this applies to all men who have believed in their own divinity or claimed to speak with the voice of God. A naturalistic reinterpretation renders all that is authentic in the Christian doctrine greater not less, for it makes it part of a new and stronger man, not of some fancied "superman," but simply man as he is but less distorted by a dissociating tradition. To rob man of his noblest faculty, *the experience of and aspiration to perfection and unity in himself*, we can now see to have been a truly hellish surgery. It took man's best away from him and led of necessity to nihilism when this belief in the pseudodivine collapsed, as sooner or later was inevitable thanks to the light of reason. For immature as reason was, it expressed the need for a deeper ordering of experience than Christianity could offer. History will be in no doubt about it: the, nihilism of the West today expresses the failure of the nondivine "light" of Christianity.

Another misapprehension must be averted. Does any reader believe that Freud's view of man held *fundamental truth*? What I have just asserted is not the Freudian interpretation of religion. Far from it. Religion, understood as the dedication to perfection, is not only healthy but necessary. It is one product of the unifying imagination, the flow into the conscious mind of the products of the unconscious, which is the source of all that is noble, beautiful, and true in man, including his intellect, which filters and scans the gifts from less conscious levels.

The author is aware that he is far from being alone in this critique of Christianity. Indeed Christianity, like the other great religions, is now everywhere in retreat, persisting mainly as a social institution, and it has become the fashion for theologians, as usual following instead of leading, for only their great originals led mankind, to write subtle dialectical essays on a new conception of God, more acceptable to a late twentieth century audience. Get onto the twentieth century bandwagon as quickly as possible! But they have not come clean, discarded the transcendental, said "mea culpa," resigned from their tarnished Church, and started afresh. (Can the shabby Christian graveyard be a forecourt of joy?) Is it absurd to expect such honesty from salaried preachers? In view of the present plight of man this is a serious matter. I consider that any true follower of that inspired but vindictive individual, Jesus, must repudiate the church, start anew, and use the understanding of our time. Some have seen this. But few Christians have realized that in this fresh self-dedication the Jesus of the Gospels cannot be their guide. For the religious persecution that has stained the history of Christianity goes back to the vindictiveness of Jesus, who threatened his opponents with everlasting fire and eternal hell.

Albert Einstein, neither a believing Jew nor a Christian, but a humane spirit seeking a universal view, wrote:

> In their struggle for the spiritual good, teachers of religion must have the stature to give up the doctrine of a personal God–that is give up that source of fear and hope which in the past placed such power in the hands of priests. In their labors they will have to avail themselves of those forces which are capable of cultivating the Good, the True, and the Beautiful in humanity itself. That is, to be sure, a more difficult but an incomparably more worthy task.[5]

How many religious teachers have had the courage to cease to be children and to follow Einstein's advice? Is the Christian religion taken seriously by its highest representatives? No; it seems that traditional good behavior in the priestly hierarchy is what counts. Otherwise why were not pope and archbishop patrolling the streets of Belfast arm in arm, day and night, until Irish religious intolerance on both sides was seen for what it is: devilish not divine? (By the time this is in print there may be some other equally ugly display of disreligion.) Of course their God would have deserted them, but their deaths might have brought a spark of nobility into the last hours of Christianity. But they had not the spunk; they are unworthy of their God. The Christian churches have neither the sense nor the piety to unite, though therein lies their only hope, for it would purify them of rubbish. Christian conflicts have gone far to make "religion" a dirty word.

The author believes that the split which Christianity imposed on man, thus producing a subhuman man without his divine faculties and a God out of this world, has now collapsed as part of the wider collapse of what he has called the "European dissociation" of man into conflicting parts. Christianity will doubtless continue as a social get together with noble music and literature offering an escape from the harsh world, but *the Christian doctrine must not survive,* and *teachers in all lands must from now on cease to mislead their pupils by imbuing them with any religious doctrine that is less than univer-*

sal. For anything less heightens conflict and is a betrayal of man in his hour of danger. This is today's moral imperative. It is now an unforgivable sin to preach sectarian Christianity. Let us clean the churches and the schools as Jesus did the temple of his God.

Humanity now requires a universal doctrine promoting diversity, and the peculiar allegorical Trinity must always have appeared as a sectarian Western diversion from the true path to those who knew the spirituality of the East. No orthodox believer of any kind can like it, but *all* the great religions are now in the pot, and a new brew is being prepared: *a universal world view promoting diversity, this time one authentic to the core.*

The rationalists of the last century and their recent complacent followers made a curious mistake, perhaps born of their conceit. They treated religion as nonsense as "opium of the people" (what a grotesque half truth!)—and the partly developed intellect, an analytical faculty, as an adequate substitute, already mature enough to take over the task of guiding mankind. This is so absurd as not to merit analysis, and its failure is seen in our nihilism. Reason is, as many thinkers have emphasized, the instrument of human desires and aspirations, and, as already indicated, not itself a source of action or power. Moreover, the value to man of religious experience—has any reader not enjoyed the experience of perfection?—needs to be strengthened by an understanding of its role in man, the organism whose supreme aim, conscious or unconscious, is to seek unity in experience—when he is not sick.

The existence of the nontranscendental religions proves that this value has nothing to do with spurious otherworld metaphysics, or delusive ideas of a personal life after death. (Who survives? The baby, the adolescent, the adult, or the old dodderer ? Priests suggest that by the grace of God it is the individual at his most mature who survives! Silly stuff—fit only for those whose failure to live here and now makes them require another chance.) English divines regarded by their colleagues as among the most enlightened are still in 1972 promising that "we shall all rejoin our loved ones," and thereby losing all credibility. But the graver weakness of Christianity is not its childish myths, but the damage it has done.

The idea of God lessens man. These words will not be forgotten, for they are both true and timely. Every false or confusing belief ultimately does harm to man, and today we are coming to understand this. Moreover, when this is known by any individual, he is on the way to being saved from nihilism, for, uninhibited by false religion, he will then open himself to anything bringing him the experience of perfection, the recovery of the joy of living, as youth today is open to new experiences, some of which may redeem. Learning the truth is always harsh to old mistakes, but we are now up to our necks in nihilism. We are suffering hell anyhow; let it be to some purpose.

It is thought by some that the world view now developing—which must become explicit and effective within a few decades if anything good in civilization is to be preserved will be based on science. As we shall see later in more detail, this is wrong. The coming world view will be more than "science" in any legitimate or traditional sense, for it must be

More than quantitative, covering order, coordination, and quality.
Absolute, in the special sense of providing a balanced union of subjective and objective.
A comprehensive view, accepted or rejected by everyone as an act of judgment, not a hypothesis to be tested by professional experiments.

The sudden development of this world view—and it must come soon and develop swiftly—as a consensus of all aspects of man, will appear miraculous to naive rationalists, if any still survive, perhaps in dark corners of the universities. But such sudden occurrences are the commonplaces of history. What may seem most novel—but wrongly, as we shall see—in this expectation is the unwilled and partly unconscious collaboration of specialist scientists in establishing their segments of a single integrated world view. The wholly natural character of this awaited historical "miracle" will be shown in chapter 7, where we shall examine a remarkable example of such unconscious scientific collaboration.

Now back to the author. His authority for writing this book is similar to the Christian authority, but more honest, less harmful, and subject to the test of history. There is here no vague dreaming of a far future. Everything asserted here will be seen to be becoming either broadly valid or not so before this century is out. This is a prophetic work, in the authentic sense of pointing to potentialities in the present. No one can today *know* tomorrow. But we can see, and live, and work for what we know to be latent in ourselves and in the condition of man today. So, finally, whence the author's judgment of the potential in man? Why has he a compelling intuitive conviction of what is valid and timely, while retaining the awareness that his personal judgments may well be wrong?

While the Christians misused messages from the unconscious to claim transcendental inspiration, the author wishes to share such messages, claiming for them an element of universality. He may not report them adequately, but that matters little, for many others will be issuing similar messages, with their own personal bias, and what is common to them all will survive. We are in a time of convergence, and many will voice different aspects of a common truth. This will inevitably occur if the race does not destroy itself first.

But what is the significance of this claim of "universality"? It means that the main ideas presented here are believed to be only to a small extent expressions of the author's ego, distorted by his experiences since 1914, such as two world wars, failures, illnesses, and so on, and to a larger extent expressions of a universal factor potentially common to the minds of all men. This "universal" is the genius of man, and its appearance in an individual will here be called *the universal.* But the character of this universal can be identified more closely. *The universal is that which reveals the optimal path toward unity.*

Every nonpathological mind seeks in some degree to order and unify experience. This is the expression at the mental level in man of the self regulating, coordinating, and formative processes of the organism. The genius of mankind, the universal in the individual, expressed in all great religion, art, and science, is the process which transcends the characteristics of the individual scarred by experience and follows the most direct path towards a unified coordination. The universal is that part of the unified truth which is timely.

The parallel between Christianity and the author's situation is instructive, but only partial. Both exploit messages from the less conscious mind in order to present a world view. But Christianity is sectarian and rests partly on a myth or allegory, while the world view presented here is potentially universal and is a fully rational philosophical interpretation of the human situation in nature, as seen today.

Any individual who feels himself called on to fill a significant role in his own time inevitably tries to understand himself and his task objectively, as part of history, in order to escape his personal limitations. All the founders of religions did so, in our century Charles de Gaulle did, and, in his own context, the author does too. But having used the Christian analogy as a vindication of his own apparent presumption and as providing a quasi historical setting for himself, the author dismisses this analogy, drops the pseudo objectivity of the third person, and becomes explicitly his conscious self, using the first person.

I am what I am today partly because my parents were what they were many decades back. My father was three parts Calvinist and one part mystic, and I have never understood how he combined them. My mother was occupied with Eastern and universal intuitions of a unity which I believe my father could never fully achieve. Two episodes throw light on their attitudes.

One day in 1907, walking near our holiday home in Invernessshire, my father, a Scottish preacher famed for his eloquence and regarded by some as a religious genius, heard the Divine Voice.[7] He was wrestling with a major issue affecting his religious calling. Should he not preach more on the gentler and more hopeful aspects of the Christian truth and less on sin and its fruits? (That my father could think this, even for a moment, endears him to me as nothing else has.) But the Voice commanded him: "Go on! Flinch not! Do that! *For no one else will do it!* [my stress]. Your appointed task is to show my people their sin." Proud man, he felt himself personally chosen by his God to demand of all that they should ceaselessly meditate on their sins. He said to his students, "I shall never forget the exact spot where that clear command came to me, where I got fresh authority and encouragement to finish my work." Authority from a transcendental God, or pride in a personal mission? ("No one else will do it"! Did not Calvin and Knox have a good try?) His vocation was to be the voice of the human sense of sin and spiritual damnation, while its primary cause was probably his shame at his own illegitimacy. No wonder that, as his biographer says, some hearts sank as they heard these words. For him the corruption of man was the primary fact, heightened by its contrast with the exquisite holiness of God.

In one respect I follow him. I believe that my inner voice is as powerful as his was, and this book is based on it. But it has told me to forgive myself for my sins and as far as possible to forget them. My pride is to be one of many who are seeking to restore to man the exquisiteness of his own highest experience and aspiration, to bring back to the individual *as of right* an awareness of perfection, of unity in thought and action. This should be Article 1 of the coming Universal Declaration of the Organic Rights of Man: *Man possesses the faculty of experiencing perfection and must not be frustrated in enjoying it.* This is the positive core of the human heritage: to be able to experience joy. But this transformation from emphasizing sin to experiencing joy is not a moral advance; it is an aesthetic/organic metamorphosis from disorder toward harmony, occurring under new circumstances. Morality is concerned with particular issues; the metamorphosis is global.

But why and how came I, his son, to emphasize so different a view? I cannot think of claiming any credit for the view which I naturally find preferable since it is my own. The question is: How did it become mine?

I have already told the story elsewhere,[8] but to omit it here would leave me obscure, for it alone explains who writes this book. My father apparently thought he had done his thing in giving my mother eight children, and he left their upbringing to her. I cannot remember ever receiving any personal guidance from him. But I do remember that when I was quite young my mother read me a passage from Henry Drummond.[9] Did she consciously will, or unconsciously design, that it should mold and guide me, her last hope? Here is the passage as I remember it, which is not quite as Drummond wrote it: "If nature be a harmony, all aspects of man—physical, organic, mental, and spiritual—fall within the circle of nature, and are inseparable from one another. There are no discontinuities between the physical, organic, mental, and spiritual."

Superb! 1883, at the time of the debate between Darwin and Christianity. All comment is superfluous, for here is the very heart of true religion and of science: no discontinuities! A great doctrine in two words, or better, a great *aim*, since no science can yet achieve it. How different from my father's obsession with sin! Could two such contrasted organisms successfully mate to produce viable offspring? Were they of one species: the prophet of sin and the prophetess of harmony?

Need I say that my mother's influence prevailed over my father's preaching? This vision of a unity transcending all our current intellectual categories physical, organic, mental, and "spiritual"—seized me and long lay quiescent in the depths of my mind awaiting its time. I am still its involuntary servant, doomed to go on cantering around seeking everywhere signs of an emerging unity, a man without a profession but to be a hunter of unity. (Who has paid me for this? The U.S.A.) The scientific search for order is one specialized expression of an earlier and more general "religious" search for unity. No psychological need for a transcendental harmonizing principle will remain when man has once seen with compelling clarity the actual ordered unity of this natural universe (to be treated in chapters 3 and 4), comprehending all its extraordinary variety and regions of apparent chaos. We are approaching a comprehensive unified view of the totality of experience, and to contribute my bit to it is my struggle and pleasure. We cannot today anticipate its full quality or all its characteristic details, but this unified philosophical view must come, and I have had a glimpse of it, seen from my own standpoint. It is human destiny to achieve this now, and this achievement like nothing else could so astonish, fascinate, and remold man as to—who knows?—even make possible the indispensable minimum of a world order. God may have died to make human unity possible. Anyhow, man will have made a wonderful bargain, swapping God for a deeper understanding of himself, his imagination, and his faculty for joy.

These words of an inadequate instrument are timely, and their source is the universal in the human imagination, authentic and liberating. I have had the fortune to know the Quaker "inner light" and something of the mystic's experience of unity, and I am not alone. Thousands have known it for every one who has put it on record. This is not a matter of esoteric privilege, as some Eastern philosophers claim, thereby manifesting a blemish in their faith.

I make no apology for these personal confessions in a book on the universe of human experience. Anything said or done which is not the expression of enthusiasm springing from the less conscious levels is without value, abortive. But personal enthusiasm must be supported by impersonal dispassion. Right or wrong, I feel the truth of what I am saying. The analytical, critical, and dispassionate intellect is indispensable, but its role is secondary and retrospective. Reason is not creative; the whole mind can be, and I am its instrument.

The fact is—and I say this in all innocence and humility as if they were my last words on this earth—that I believe myself, like many others, to be the inadequate and probably biased carrier of universal ideas which have over several decades come to me as free gifts from the less conscious levels of my mind. Believe me, I seem to be strangely constructed, though Einstein felt the same, Blake said of a major poem, "I can praise it since I dare not pretend to be other than its Secretary," and Rimbaud that "I am present at the birth of my thought, I look at it and listen." This sanctions my confidence in offering you these pages. There is no ground for surprise at their wide coverage; they are written by a generalist who holds, as he believes, a clue to unity in many realms.

The central ideas here are in no sense mine. It is true that having arisen in my trained mind, they are, or can be made, as precise and powerful as I could wish them to be. But I do not feel them to be mine; I possess and desire no monopoly, or priority, or privilege through them. To seek that would be to renounce something much more precious: the awareness that one is part of a universal movement, glorious, free, and open—much grander than one's ego! I once read with a shock what a distinguished scientist, Peter Medawar (I don't believe in using titles when they are not appropriate), wrote with his knowledgeable scorn of scientists' motives. "Of course everyone is out for himself, and wants to be first." *Everyone?* He must be unaware that some tasks are so big that one wants others to share the burden or even to take the first step. But it is a good thing we are not all alike. Philosophy, science, art, and the human community need all kinds. Ambition, poverty, and fear must play a role, as well as talents, imagination, and the bliss of sharing in a great adventure. I see no point in considering one motive as "good" and another as "bad"; many motives are unconscious, and moral comparisons are empty. Things are as they are, and individuals vary, thank goodness. At my present age I do not even care much whether these ideas are right or not, for they can be displaced only by even more beautiful ones. If I have deceived myself, it has been a grand dream.

In a more precise sense these ideas are not "mine." I did not *will* their formation, any more than I can will my dreams, nor can I will them to let me be. Just as I had no freedom of choice but to accept them whenever and wherever they came to me, empowered as they were by intimations of great promise, so I have no choice but to go on and on trying to share them with others, over and over again, as they gradually acquire greater clarity, strength; scope, and timeliness. Their mark as they announce their coming arrival is the distant music of a promised beauty: "You cannot help attending to me, you already feel how lovely I am, let my beauty enter your awareness."

They promise an exquisite experience, as perfect as "the holiness of God" was to my father. As a unifying idea approaches, it is sheer delight to be in its vicinity.

Moreover, these ideas make on me a strange demand to which I must accede. They remind me of what Camus called "having the courage of one's own good feelings." One must not be ashamed of admitting that sometimes one possesses a powerful source of joy. In an age of nihilism one must not be shy of standing up to be counted. "Look! Another would be scientist gone religious!" No! For such terms have changed their meanings. I look beneath science and religion to the human imagination that created both; my task as an interpreter is to try to look ahead; and at long last I am no longer ashamed of displaying my convictions without equivocation.

It is part of my confession to stress that a point can be reached where personal inadequacy for such a task does not matter: This work may be my swan song on universal themes, and I know that it is imperfect, but it is the best I can do. The music. that I have heard may come through. If it is indeed universal it must sooner or later, and that is all that matters. My words will acquire an eloquence which is not their own in the minds of those to whom they are congenial, as beauty is in the eye of the beholder. Men tend to dislike those who worship a different God, but they love those who show them their own. Then the music comes through and makes us one in a communion which is neither mystical nor erotic, but of a kind for which there is no name.

In this gamble I am rewarded by surprises. The convictions I possess are stronger than my ego! They have on occasion actually helped me to live, exercising a therapeutic power, though I will confess that at times they have deserted me. They compensate me for the loss of Christian faith. Indeed they express a quality which no ascetic male saint—think of Gandhi—tortured by the flesh could possibly sustain without moments of terrible failure when the devil laughs at his hypocritical piety. Moreover, they command the services of my analytical intellect because they seem to me to be born of a deep necessity. This is the way the human mind must go in this century, the only sane way toward unity. There is nothing arbitrary or contingent about them. I repeat, if a unified world view is to be possible they are necessary.

There is nothing vague about this uniqueness and superiority. This is so important that I will set out here at once the manner in which these ideas are distinguished from all alternatives conceivable to me in the 1970s. Here are their compelling marks. At this point I anticipate the theme of chapter 4 so that the reader may know where he is being taken.

The ideas presented here concern the entire universe of human experience; they are latent in all experience and throughout nature. The term "nature" is here used for "the total universe of experience."

They are absolute, as already suggested, in the sense of being both subjective and objective. They link subject and object, for they express a tendency in the mind which enables it to identify the same tendency in nature, including the human brain.

They are aesthetic and philosophical, but *post*religious in the sense of discarding transcendental divinity and *pre*scientific, expressing ideas not yet accepted by the sciences, though they must soon be seen to provide a basis for deriving particular scientific hypotheses.

They possess immediacy, being directly based on our experience of three dimensional space (3D) and of temporal succession (S).

They possess great power,[10] and must provide the unifying principles of the next synthesis of physical and of biological theory. They are fundamental, and I believe that they can cover the content and reduce the complexity of every situation, by revealing latent order.

They are organically therapeutic, easing the return to health after physical or mental disturbance of the individual who accepts them, and possibly even aiding the creation of a world community after the age of nihilism, though that is only a personal hope. *But they give no guarantee whatsoever of recovery or success in any personal situation*. They evade utilitarian evaluation.

They are timely, because they grew in the less conscious realms of one mind, and perhaps of many, as an organic response to the hell of 1914 onwards. They are urgently needed by all, young and old, male and female, white and colored, everywhere.

They possess a potentially universal appeal because they spring from universal factors in the unconscious levels and are not tied to any local or sectarian tradition.

Moreover they spring from a single conviction which is simple and powerful once its implications begin to be seen:

> There is present on many levels in nature a tendency toward order, form, and symmetry; hence in living systems toward organic coordination; and in man toward personal coordination; this tendency being realized when circumstances ar favorable.

This is a first glimpse of the world view which I present in chapter 4.

To some readers this conviction may appear to be no more than an unproven but interesting intellectual conjecture. To others it may appear as an expression of aesthetic faith. Yet when its fertility is recognized by the sciences this conception of a pervasive order generating tendency will be seen to be one of the most powerful ideas which the conscious human mind has ever received from its less conscious levels. It is not new; consciously or unconsciously it has haunted many civilizations. But in our context it has a new significance and power. When it has seized the whole person it becomes more than an intellectual idea. It implies a mode of feeling, a style of thinking, and a way of living, a Tao in words. If I am not mistaken it lies behind all past religions and present sciences. It is the aim of these pages to show that this idea is rooted in the character of the entire universe of our experience. The time for this universal principle has arrived, and it must now unfold in the minds of many. Where I have added personal ideas of which I am uncertain they are indicated; the universal and my hunches must not be confused. Anyhow there is enough truth here to make my personal errors obvious and harmless.

This book is the result of a decision to wait no longer for exact science to confirm my identification of the hierarchy of formative processes and my recognition of its scientific power, but to go ahead and put it on record while the going is good.

Changes of opinion and judgment occur with startling suddenness when the appropriate moment has arrived. This interesting fact will be considered later. Here I merely assert as a historical prediction that we are near such a moment, or perhaps already in the middle of a period, when a relatively sudden metamorphosis of the psyche is occurring, bringing into full awareness what has been slowly developing on less conscious levels since 1914. This implies that the ordering tendency, here called *morphic* (see chapter 3) will before the end of this century be taught to children in many lands as one of the axioms of a new world view. This is my prediction, made in 1972 and placed here on record. Beauty is no guarantee of truth, certainly, but so elegant an idea is worth trying out. In chapter 3 I shall develop the background of the world view. But first we must examine the position of science today.

2

Science, Scientists, and the Universe

Incomplete sciences mislead

It is said that science is on trial. Yet there is no fundamental separable entity, "science." Our complex culture is pervaded in varying degrees by components which may be regarded as "scientific," but there is no agreement on the precise meaning of this term. This caution regarding the term is not verbal pedantry; it is a matter of urgent social importance.

In most cases where the term "science" is used, no harm is done, since the context shows what is meant. But on fundamental issues the term "science" is already creating confusion, and this will grow worse. Science originally grew out of the religious and intellectual search for order and unity in experience, as a branch of culture using quantitative observations and theory. Now, partly as a result of their successes, the various sciences have to be regarded as parts of a wider world view the basis of which is not scientific in the 'strict or traditional sense and is better regarded as philosophical. The traditional separation of science from the rest of culture was a superficial and temporarily permissible procedure which must now be dropped when fundamental issues are being treated. It is the unprecedented transformation of the conditions of human life by what is called "science" that renders greater clarity necessary.

That the common and vague use of the term is basically misleading is beyond question:

1. There is no agreed and permanent boundary between philosophy and science; the supposed boundary continually changes and does not apply to the life sciences, though this may be controversial.

2. There is no one differentiated form of activity that can properly be called science, for there is no unique scientific method, as the positivists wrongly assumed.

3. "Scientists" do not display any characteristics which distinguish them from the rest of mankind. Enthusiasts intuitively searching for an assumed theoretical unity are as necessary as design engineers.

4. The assumed objectivity[1] of certain scientific methods, though successfully maintained in some realms, has become restricted and conditional in others. What we regard as "facts" depends on our approach. In many fields the experimenter by designing the apparatus in a certain manner selects what will be observed.

5. "Measurement" is no longer a clear idea, for quantum mechanics lacks a satisfactory theory of measurement, and in some realms it may be order rather than quantity which counts.

6. It is now widely understood that the special procedures of science rest on a background of tacit knowledge[2] and concealed metaphysics.

7. As science approaches man it ceases to be value free. Some cling to the separation of fact and value, but we shall see that a science of order necessarily includes values.

8. Finally, the existence of human consciousness may require for its understanding an approach unlike those of any science hitherto. The solution of the body/mind problem may require a new philosophy.

These points require further analysis, but their cumulative effect is decisive: a new clarity is needed. Most will agree that the unmistakable objectivity at certain levels and within certain restrictions of some classical scientific methods has emancipated man from intellectual darkness, and must be retained and developed, *probably with a new significance.* The interpretation of science as an unbounded region within a new comprehensive culture will, we must hope, lead to a deeper understanding of the meaning, scope, and value of scientific objectivity, perhaps replacing it by some different but equally reliable guarantee of truth.

These issues appear timely, indeed urgent, because this is "the age of science." Yet even this term is misleading. The period from 1600 to date has been one of partial sciences lacking basic theoretical clarity and unity and therefore necessarily unreliable, since somewhere they must contain mistaken assumptions. As Henri Poincaré said in *1913,*[3] *"One should be fearful only of an* incomplete science *which deceives itself,"* and us! In my view in the late twentieth century we are passing from the age of partial and therefore deceptive sciences into that of a universal view within which the various unified sciences will find their place as components of something more fundamental: a general philosophical doctrine of order and disorder covering both the objective and the subjective aspects of experience. In the advance toward greater clarity a new austerity and self discipline is required of scientific specialists and we must hope will be accepted by them because they are, or should be, first humane individuals, and only secondly professional scientists.

The fact that incomplete and only partially unified sciences must somewhere prove misleading draws our attention to a remarkable fact which though obvious to specialists, or perhaps because it is obvious, has not yet received any explanation or interpretation. *After nearly four* centuries *of quantitative research no exact science has yet attained basic theoretical clarity!* This is remarkable and merits attention. We can send men to the moon and back without fundamental intellectual understanding of any realm of nature! Of course if science were an endless search, and knowledge always open ended, this lack of definitive clarity would have to be taken for granted as one of the unavoidable facts of life.

But this interpretation may be wrong, and it is certainly not productive. It may therefore be useful to consider another possibility. The facts are beyond dispute. Physical theory lacks comprehensive unity.[4] Biological theory is still at an early stage and lacks clarity.[5] Psychological theory and social theory are in large degree matters of personal preference, as there is no agreement on basic principles. The extraordinary fact—that science nearly four centuries after Galileo and Kepler has nowhere yet achieved basic theoretical clarity—may have a discoverable meaning. Or to put it another way: What does it signify that science has reached a stage where scientists can know beyond any doubt that all existing theories are merely incomplete and faulty approximations? In 1930 a few naive theoretical physicists for a brief period asserted that quantum mechanics was the final word on the physical universe, though Heisenberg and Einstein did not make that mistake. Few would maintain that today.

Yet the aim of intellectual clarity is not merely a dream, for in restricted respects mathematics has achieved a satisfying clarity. Is a similar clarity about nature beyond our reach? Our ability to pose this question today is open to a constructive interpretation; the clear recognition of a problem is often the prelude to its solution. Science may today be on the eve of a new clarity. After a period marked by the collection of an immense volume of relatively disordered facts it is natural to expect what has often happened in the past: the formulation of new clarifying and unifying principles. To be aware of a pressing need for new principles is to have passed over a threshold, and disciplined conjectures about the way ahead can then serve a purpose. My conjecture involves the assumption that the next comprehensive clarity will take the form of what must be regarded as a philosophical world view subsuming all the special sciences. The various sciences now become the instruments of a type of man marked by a new basic awareness expressed in a philosophical world view appropriate to the late twentieth century.

This raises an awkward, indeed painful, question. Can scientific specialists be trusted? When, if ever, is a scientific orthodoxy reliable?[6] Five decades spent in watching science have given me the outsider's advantage. He can sometimes see general features more easily than those continually occupied with special professional tasks. One thing I have learned with amazement, as I began with naively favorable judgments and respected my seniors, is that *nearly all scientific specialists exaggerate what is already known*. This is an occupational disease to which in principle all are vulnerable and those who escape it are rare. The Cavendish Professor of Physics tells us (1971) that "a physics research student is apt to overestimate his understanding of things in general. . . . This unconscious arrogance is not confined to research students." It has been said that premature generalization from limited evidence is a weakness of most psychologists. That is certainly true. But it is equally true, though perhaps less obvious, that in all realms specialists tend to mistake the assumptions of some temporarily dominant theory for ultimate laws of nature. Young skeptical students tend to become dogmatic professors. It seems that you cannot earn your living by teaching a theory the validity of which is restricted without unconsciously becoming identified with its dogmas. What would happen to a university if every teacher of science sought only to promote truth and spent the first five minutes of every hour saying, as he should: "What I am going to tell you this morning must be wrong, since we do not yet possess a unified theory of this realm. There must exist unconscious, redundant, and faulty assumptions somewhere, but no one yet knows where"? The scandal would spread, the shaky foundations of knowledge be exposed, fall outs multiply, and skepticism result in nihilism, which is where we are now. Something must be wrong in this vicious spiral of uncertainty.

Do not imagine that I am merely being ironic, a devil's advocate of being conscious of what is by definition unconscious. For it is not always difficult, indeed, it is sometimes rather easy, to discover what have previously been your own unconscious assumptions. You smoke them out, forcing them into the open. The first step is to know that they must exist. If you are honest with yourself, and really believe this, the battle is half won. The next step can be fairly easy, if you use a psychological trick. Pretend that you have an academic adversary, if you haven't one already. Then ask yourself: What more general assumptions would *he* make if he wanted to displace your theory? Use him as your alter ego: make your image of him start on the job. Then decide to identify the limiting flaw before he does. After this, relax, contemplate the problem every day for weeks or months, until one morning, Snap! You have the answer. This nearly always works. Properly nursed and teased, the unconscious levels can be very helpful.

My knowledge of the history of philosophy is scanty, but I know of no systematic philosopher in modern times who applied this method of ironic self criticism to his vaunted doctrine, as every lover of truth should. Did Bertrand Russell, a great thinker and a great man, ever tell his readers in what kind of universe, or for what kind of man, his doctrine of "logical atomism"—the assumption that the world consists of simple elements, such as percepts or sense data—would be false? I do not think he did. (This assumption is not appropriate for a theory in which man is not separated from the external world.) But Wittgenstein, in his persistent self communing, tried to do precisely that and as a consequence discarded the emphasis of his first period and developed that of the second, in which he laid stress on the historical approach rather than on logical clarity. Yet his long and lonely journey was not necessary, for if he had introspected less and read more widely he would have learned what many thinkers already knew: that the historical approach was in this century becoming more powerful than the analytical, which in some realms had been overdone.

Everything worth saying can be put in a few words. What I have just said can be reduced to one word of three letters: TAG. I suggest that by edict of the American Association for the Advancement of Science or of the Royal Society, from now on every generalization published by a person claiming to be a theoretical scientist should carry a tag stating what the author had by the date of submission discovered about his previously unconscious assumptions. If the editors of scientific journals care about truth, this ruling will be put into force tomorrow. Whitehead defined in five words a general policy for science with this warning tag: "Seek simplicity, and distrust it."

If individual specialists tend to exaggerate their knowledge, this personal weakness becomes a social scandal in scientific orthodoxies. This is an old story, but it has recently become more serious. For the current professionalization of science, the grouping of scientists into large institutions with hierarchies of leaders and followers, has made the role of scientific authority greater than ever before.' Planck said that "a new scientific truth does not triumph by convincing its opponents and making them see the light, but rather because its opponents eventually die, and a new generation grows up that is familiar with it." In some realms Planck's generalization is exaggerated, but the inertia of science has now acquired a new form in the weight of the great governmental and private research institutions. This must be compensated by teaching a new skepticism. Arrogant orthodoxies must be shamed out of their misbehaviors. *Tease them night and day about the tag.* Keep asking them: Under what conditions are your assumptions valid? Your mathematics may be pretty, but what is its precise range of validity? Where do your concepts and assumptions break down? This will annoy them, but one in five may feel challenged to try to answer you.

Only the elderly or old can know the experience of seeing young skeptics becoming ruthless dogmatists as they grow old and secure. I have found it depressing to watch through a life time the slow emergence of a fertile idea, its supporters risking everything to promote it, and its subsequent congealing into an arrogant orthodoxy, some of those same men taking their revenge and punishing by deliberate neglect those who dare to challenge them. I know from my own experience that an army in battle (old style, 1914–1918) displays more courage, at the lower levels at least, say from captains downwards, than most academics do. But fortunately the more dogmatic an orthodoxy is today, the more likely is its collapse tomorrow. After twenty or thirty years of unconditional dogmatism the breakdown may be close at hand, and often silent figures on the sidelines have all the time known where it was false and have waited for the moment of truth. The best scientists are not necessarily the noisiest. Orthodoxies neglect the history of science and *never learn humility and caution.*

"All genetic determinants are in the chromosomes." This glib assertion had quite a run, in spite of its being improbable on general biological grounds, for no differentiation is absolute, and botanists knew it was wrong. It is now known that there exist cytoplasmic determinants, and perhaps the next discovery will be that the cell cortex (ectoplasm) plays an indispensable role in heredity, as of course it must, though probably not as the determinant of specificity.

But here is a more important and subtle example. Watch it; it will be fun! Under the neo-Darwinian or synthetic theory of evolution, which has been dominant for some thirty years, "random mutations" provide the opportunity for natural selection to operate. But some of those who have become aware of the highly structured character of the living cell, from the cell itself down to the biomolecules, consider that this doctrine may be misleading, since it neglects the role of this internal structure in molding random *changes,* say produced by radiation, into *mutations* which are stable enough to serve as hereditary determinants." I expect this view to be widely accepted soon. But it is unlikely that representatives of the orthodoxy will cry: *"Mea culpa!* I was wrong!" It is more likely that it will be left to a younger generation to correct their errors. (In the West we do not believe in putting men in prison, or even withdrawing their pensions, for teaching what is misleading or mistaken, and a good thing too!)

I once started collecting material for a satirical history of science showing the blunders made, the supposedly crucial experiments which actually proved little or something quite different, the apparently slapdash methods of genius, the persistent use of unclarified terms, and the inadequacy of much of the dominant "philosophy of science." For there cannot exist one "philosophy of science" when there are so many contrasted sciences using different methods which are anyhow always changing. One task of that penetrating history would be to explain how science has managed to advance in spite of the bizarre behavior of so many scientists. But I have been kept busy on less entertaining tasks.

All I have room for here is the blunt assertion that if the past history of science is any guide, some highly distinguished professors are talking nonsense at this very moment, which is hardly news. There is nothing surprising in this; as we shall see in a moment the very greatest, such as Maxwell and Einstein, can make strange blunders, being, like all of us, men of their time, even if a step ahead. The reverence for great names has gone too far and damages science. The coming reaction from the sentimental adoration of Einstein is going to be more violent than it need have been. (His Special Theory cannot play a part in a fundamental unified theory without reinterpretation. See chapter 3, note 7.)

The human mind, and the scientific intellect above all, seeks order and simplicity. So when a simple generalization seems to meet the facts the weary mind believes it can relax at last, and the alternatives are inhibited from consciousness. Here an interesting psychological mechanism may come into play. Unaware of what he is doing the author often overemphasizes the hasty generalization, the alternative being inhibited by a redundant emphasis, as we shall see.

It is when scientific specialists make unconditional assertions about the universe as a whole, or about local principles applying everywhere, that this common frailty becomes most evident, and most dangerous. Here the temptation to slip into absolute assertions is too much for even the greatest to resist. The desire for a simple clarity applicable everywhere, the aim of the mind's search for order, here becomes highly treacherous. The simplicity, clarity, and certainty which are the aim of all searching must be the characteristics of the theory already formulated.

Some of the greatest—I have Einstein and Dirac in mind—have suggested that for the theoretician elegance (used of course by men like them!) may be the supreme criterion of truth, prior even to observation or experiment. For a period I also indulged the fancy that the highest mathematical elegance was a better guarantee of truth than any one experiment, which might always be faulty or misinterpreted. But this primary emphasis on elegance is mistaken, and dangerous. *Theoretical elegance cannot guarantee truth*, for in a decade a fertile mind can generate several superb ideas each of which may for a period strike him as supremely beautiful, though it may be that only one at most can be valid. Kant was right when he suggested that the sense of beauty is an aid to the discovery of truth. An aid, yes; but a sufficient criterion, no! There is no alternative to ceaseless searching and irony toward what for a period may seem to be successful and valid. *Even the greatest minds do not merit our trust.* The greater their reputation and the more we admire them, the more treacherous it may prove to be, if one sentimentalizes them into father figures who cannot err.

This recognition is crucial for society today. If a philosophical or metascientific world view has now to be developed, the highest scientific generalizations must find their place in it. But who can be trusted? Consider the following blunders made by noble minds, whose merited reputations cannot be damaged by my pointing out one mistake. Homer nods; Maxwell and Einstein slip. Scientific genius always displays what is felt by contemporaries to be a naive belief in the possibility of great simplifications. The price for this belief is that great scientists can make great mistakes.

1. In 1875, James Clerk Maxwell wrote:

> We cannot conceive any further explanation to be either necessary, desirable, or possible, for as soon as we know what is meant by the words *configuration, motion, mass,* and *force,* we see that the ideas which they represent are so elementary that they cannot be explained by anything else.[9]

These words were written at the high tide of the mechanical philosophy, when Maxwell was himself searching for a mechanical source of his electromagnetic quantities. I know of no more superb example of a highly talented figure expressing so clearly what were soon to prove the mistaken assumptions of his time. The historical irony is complete. For Einstein, who deeply admired and loved Maxwell, spent a great part of his life showing that here Maxwell was wrong, that the concepts of motion, mass, and force could no longer be regarded as primary. The triple emphasis, "necessary, desirable, or possible," indicates Maxwell's complete conscious satisfaction with the mechanical basis of physical theory. Did he, at some deeply inhibited level, faintly suspect that his electromagnetic quantities might rival the mechanical concepts and be equally fundamental? We cannot know. But this example shows that no past genius, however great, is to be trusted; our historical situation is different from his.

2. We now move on thirty years: In 1905 Albert Einstein, then only twenty six, wrote:

> . . . All our judgements in which time plays a part are always judgements of **SIMULTANEOUS EVENTS** [his stress]. [10]

This is incorrect. Physics also continually requires the judgment, firstly, of a time interval having passed between two events—which Einstein took for granted—and, secondly, and more important, of a "later than" relation between them, as physics has to deal with many processes displaying "time's arrow" (see below, and chapter 3, note 7). Einstein's attention was temporarily focused mainly on simultaneous events. A single slip in that prodigiously productive year 1905 is not surprising in one so young, but many thousands of readers of that famous paper have treated it as gospel truth and have been misled. They took genius as faultless, and failed to notice the mistake. Even Aristotle, if we may trust his translators, knew better, for he wrote: "It is only when we have perceived 'before' and 'after' in motion that we can say that time has elapsed."[11] Notice Einstein's "all," "always," possibly indicating a deeply inhibited doubt.

3. We now move on another twenty three years to note a similar error, this time by a highly talented figure of lesser stature, whose interpretations of physics exerted a great influence on teachers and students. For that reason his mistake has laid a curse on at least two generations of physicists.

In 1928, Arthur Stanley Eddington wrote:

> So far as physics is concerned, time's arrow is the property of entropy alone.[12]

(On entropy see chapter 3.)

This is incorrect today, and was so then. There are many other kinds of physical processes which display "time's arrow," that is, a one way tendency.[13] To take one example, an excited atom possessing surplus energy radiates spontaneously, but an unexcited atom does not absorb spontaneously, that is, without receiving energy from some source. What Eddington should have said (and perhaps meant to say?) is: "The only general law yet discovered which displays time's arrow is the entropy law, but of course there are many physical systems which do so." Note the similarity to Einstein's mistake above. They both underestimated the importance of the relation "later than" in physics.

I have been surprised at the persistence of Eddington's error. In 1965 A. L. Lehninger, a distinguished specialist on bioenergetics wrote in a justly popular American paperback: "It is a *fundamental* law of thermo-dynamics that *all* atoms and molecules in the universe *inexorably* tend to seek the most random or distorted state."[14] Sorry! It is not! His assertion is true only of closed systems at a certain level of analysis. (See chapter 3, note 3.) (My italics indicate the remarkable triple emphasis in asserting a falsehood.) The author courteously accepted my correction, but he has not advised me that his slip has been corrected, and many thousands of students must have innocently absorbed this mistake, which even today may still be virulent, corrupting the minds of youth. Specialists who make this mistake are confusing what they happen to know about general laws concerning a particular class of systems with the facts of the objective universe.

I have given these three examples to show how careful the community must be today in accepting advice from scientific specialists. The lesson is that specialists should always be more careful and expect surprises, above all in their own realm, where they tend to dogmatize. Only the passage of time, and criticism by their colleagues or by others less specialized, can eliminate specialists' errors. Above all, *high generalizations are always suspect.*

From every side, even from most unexpected quarters, evidence pours in of the danger of trusting specialists.[15] Here is a disturbing example. *The Bulletin of Atomic Scientists* (Chicago) is in its twenty-ninth year, and is one of the journals of which the United States can be most proud. But its distinguished co-founderand editor in chief, the late Dr. E. Rabinowitch, a humane and enlightened person, used the following argument in favor of the maintenance of intensified research. I am loyal to his thought in condensing it thus:

1. Human reactions are still primarily emotional and qualitative.

2. We need less emotion and more quantitative research, in order that:

3. Quantitative knowledge may be adequate to guide social policy.

I am unhappy that a fine mind should believe that emotions are mainly bad, and that quantitative rules alone can and should guide mankind. Is this the doctrine for which the *Bulletin* and its distinguished sponsors stand? Is this the highest enlightened mind of the United States? I cannot believe that all the dedicated work that has gone into the *Bulletin* over nearly thirty years should now be associated with this hellish doctrine: Mankind is to be guided solely by quantitative knowledge! I plead with those responsible. *Have this editorially corrected.* This blunder is far from unique. I have just read in a scientific journal, "What the world needs is more common sense, science and technology." Whose common sense? Where do values come in?

At one time I considered that it was an intellectualist error to assign great importance to the role of ideas in social history, relative to what I then felt to be deeper organic and social factors. This I now see was wrong. Ideas serve to develop and stabilize particular human modes. New ideas can transform the historical expression of man's hereditarily determined potentialities. Each type of man selects the ideas which meet his needs, and ideas thus selected can be, and indeed often are, of immense importance. Human communities require appropriate facilitating ideas. Today nihilistic parents and their children in rebellion against a nihilistic society can be powerfully influenced by ideas indicating a sane line of advance. Indeed the ideas presented here may be among the many catalysts and stabilizers of a metamorphosis of the psyche leading toward a sane human community. Reader, if you already possess a world view with which you are satisfied, read no further. It is the contemporary vacuum that forces me to try to share my convictions.

The coming world view cannot be built on technical science, as I have shown. It must be a wider outlook which can only be called philosophical, comprehending and using the special sciences. If I have not already sufficiently demonstrated the folly of uncritical reliance on the specialized sciences, here is another argument, for me of great importance.

I assume as the supreme criterion of a humane society *the quality of individual experience.* The human subject is more important than anything else. All the greatest doctrines, religious or political, have found their ultimate sanction in the quality of personal experience and aspiration, for in this lies the dignity and uniqueness of man and his ability to transcend himself. This is the treasure we must restore and preserve. Any doctrine which seeks to override or neglect this is the enemy of our humanity. But unfortunately many scientists escape from considering the self by concentrating on external observations.

There is no need at this stage to examine in detail what kind of "quality" is desirable. It certainly must imply more than most now understand by "happiness," "fulfillment," or even "self realization." These terms are corrupted in the context of our declining civilization, which has no words left for authentic quality unless we expressly put our sense of quality into the terms beauty, truth, and above all joy. Why has joy been so much neglected? It may be the supreme value, prior to all others. The joy which passeth understanding.

Yet by a strange irony which should be frustrating to those who put their entire faith in science, if any do so still, no science, and no group of sciences, has yet any understanding of the status or role of "consciousness" in man, in animals, or in the universe as a whole. My view is that until we can interpret mental processes as organized in a hierarchy of levels, each marked by its function, we shall achieve no clarification of this puzzling issue: the status of consciousness in the universe. (See chapter 7.) Yet if the quality of awareness is primary this points to a serious lack incontemporary science. We must clearly be on our guard. A science cannot be regarded as objective which neglects a feature of primary importance.

It is to prevent this lack from being forgotten that I have chosen as my title *The Universe of Experience,* and from now on I use "nature" to include the mind. I am concerned with the totality of immediate human awareness, and all its expressions: emotional states, intellectual processes, scientific observations, and all that can be inferred from them. To begin by separating the physical universe from human consciousness certainly distorts our thinking. The existence of the self, capable of awareness, must form an integral part of any total world view. The universe within and the universe without are two aspects of one total universe of experience. But we should not fall into a "deep subjectivity," as has been suggested! We need balance.

The apparent paradox of "consciousness" in a "material" universe may prove to be the result of a mistaken approach which tends still to assume two substances or modes of being: "matter" and "mind." Yet the physicist is aware that the mechanical view of the natural universe has for a century proved inadequate.[17] Matter has been dropped, but what has been put in its place? Terms such as structure, pattern, field, and interactions are now in use, but wherein lies their essence? That the answer is difficult is due to the fact that for a century, since 1870/80 to be precise, basic physical concepts have been becoming more and more abstract, involving a steadily increasing number of logical or mathematical steps away from what is immediately given: 3D and S. It is not surprising therefore that on basic philosophical issues current physics is strangely silent, and that when it seeks to say something speaks with conflicting voices.

There is another reason for caution. The most mature and fundamental of the sciences no longer knows what it is doing! "Measuring," of course, but what precisely is meant by that in terms of the latest theories? Physicists do not know! Theoretical physics does not know how the numbers which constitute its essence come into existence! Current physical theory is thus incomplete and so must be misleading, quite apart from its other well-known limitations. This warning should not be neglected.

That a new start is necessary can hardly be doubted. Stand back and take a cold look at human knowledge and ignorance today. I remember the shock I experienced when I first realized that *we possess no fundamental knowledge whatever.* Like Socrates the only thing we can be certain of is our basic ignorance. I mean this in the most literal sense. Centuries of philosophy and science have left us fundamental ignoramuses. No reliable assertion can be made today on any fundamental problem, except this assertion of ignorance. None of the sciences is unified, not even separately. Gödel has warned us that algebra is impotent to provide a manifestly self consistent theory including arithmetic. Biology possesses no fully reliable ideas. Psychology stutters. What has gone wrong? Where are we?

The greatest advances in thought do not result from further grand generalizations, or from formal manipulations, but from the identification of some specific, relatively new idea which happens to be fertile and timely. I make the assertion which is not mine, but a gift from the universal: the failure of scientific thought to obtain more clarity on fundamentals is due to its neglect of asymmetrical relations, such as temporal succession. (The meaning of this term will be explained in chapter 3.) This is the deepest intellectual insight of which man is capable in 1972. I cannot qualify what I know to be true. This judgment has come to me from the deepest level of my mind, but it satisfies all levels from the imaginative sources to the conscious critical intellect. Hence its power, and my certainty.

Man, the thinking and speaking organism, is today a nihilist, wandering without map, compass, or aim because of this neglect. There is no guarantee of the reliability of any science until it is unified, and none is today. It may be that immediacy, the return to ordinary space and the experience of succession, must be our guide. The question is: What do we immediately observe?

3

Disorder. Order, and Hierarchy

The universe is the arena of a contest between two apparent antagonists: the tendencies toward order and disorder. Yet the ordering tendency predominates and may contain the other.

"God! She's beautiful!" I thought. "Again that dazzling blend of Aryan and Oriental." It was some years ago at San Francisco airport, and my senses had been heightened by California. I wanted my emotion and thought to lead to action, but that was impossible. Action being frustrated, introspection took its place, and this episode became a parable of deep convictions.

In this world of conflict, a beauty is possible that can bring, even if only for a moment, a sense of perfection. In principle no other justification of human existence is necessary. A moment of beauty is potentially eternal, and can restore us to ourselves by reminding us that the experience of perfection is always possible.

Yet action had been frustrated, and one cannot base a life on one glimpse of perfection. So I reached this thought. How much it would mean if the human psyche could be so ordered that its roots in feeling, thought, and action were united in a single response! Could we not reconstruct our way of thinking by starting from the point at which feeling, thought, and action are not yet differentiated? Start with *beauty,* perhaps? Hardly; that is too indefinite. But why not with *order?* For some form of order underlies nearly all human experience. The universe of experience is a cross of order and apparent disorder. Our way of thinking today is disordered and confused, and hence the present challenge to mankind evokes no effective response. We are rich in facts, but poor in ideas, and so inept in action. Imaginative leadership is not possible without unquestioned convictions founded on some passion such as the love of order or beauty.

Is there an anomaly here? We need passionate convictions, clear ideas, and effective action. But are passion and intellectual clarity compatible? The question displays our pathology. Because we are sick, we imagine that passion and logic cannot blend. (Russell, more profoundly than any other personally known to me, strove to combine a passionate, indeed mystical love with intellectual clarity and appropriate action. But he found this difficult, and to others it seems paradoxical.) Yet the mark of Unitary Man—as I have called the more harmonious individuals—precisely passionate convictions guided by clear ideas into effective action. Enthusiasm and reason are then allies. Our passionate convictions need not be rationalizations of self interest, as Marx suggested, nor, following Freud, mere idealizations of frustrated instinct. We discard the transcendental God, not in order to achieve absolute harmony, but to lessen the present disharmony of emotion, thought, and action, by discovering the source from which all three spring.

But lost in confusion and conflict as we are, where and how can we start afresh? Here I long for an eloquence to let my words sing in your mind as the idea I now present does in mine. It is the inner voice offering what is truly lovely, sheer balm to the weary human psyche, and no illusion. The universe is more ordered than our minds are, and more ordered than was recently believed. *There is more order in the universe than there is today in human thought.* Having paid too little attention to order, we must now take a hint from nature, from the world of experience as a whole, and adapt our thinking better to nature and to our deeper selves. As A. B. Johnson said in 1828, we should "subordinate language to nature, making nature the expositor of words, instead of making words the expositor of nature." Most of the followers of Marx and of Freud made the mistake of using their leader's theory to interpret facts, instead of using the facts to improve the theory.

"Mechanism" is not the element common to external nature and to thought; an ordering tendency is. So adapting our thought to nature without and within, we replace God by the nisus toward order evident within us and around us. This is an old theme, at last to be made fertile. A neglected principle of order, or better, a process of ordering, runs through all levels; the universe displays a tendency toward order, which I have called morphic (see below); in the viable organism this morphic tendency becomes the tendency toward organic coordination (not yet understood), and in the healthy human mind it becomes the search for unity which gave rise to religion, art, philosophy, and the sciences. There is my argument in brief. We must discover by trying it out whether the ordering tendency in organic nature and in the human organism and mind, properly exploited, is not more helpful than God ever was. If only simple ideas can save the world, here is one that might. It has great power.

But what is "order"? Definitions are useful, so I will give two, though we must remember that the effective meaning of the word comes from the use to which we put it. A mathematician might say: "Order is a determinate system of relations," a single system determined by some rule, either already intellectually known or intuitively apprehended, or possibly still unidentified (in which case it becomes apparent disorder; see below). The relations may be spatial, temporal, numerical, or social. In this work the emphasis is on spatial relations in that ordinary visual space in which we act out our lives and experience temporal succession. What is crucial is that being human we are predestined to take sides, to love beauty and appropriate order, and to hate excessive disorder, unless we have become pathological, like Antiman. We do not need to be taught by parent or teacher that beauty appeals to us. That bias is in our heredity as organisms. But order is a general philosophical term which becomes scientific only when a special type of order is selected and defined.

"Disorder" is subtler, being negative, and is perhaps really an expression of ignorance. The mathematician may say: "A system is called disordered when we cannot find in it some expected type of order." Each case must be considered separately, perhaps over many decades, before we can feel confident that we understand it. "Disorder" may prove to be, like "random," in a fundamental analysis a self contradictory concept. Concepts which imply an element of ignorance may at any time become irrelevant. Until now disorder has been treated only statistically,[1] for example in an assembly of units subject to random motions, whereas order marks single units.

When we gaze at the starry sky on a clear night we experience awe. It is beautiful, but at a less conscious level uncomfortably challenging. Our mind is held in suspense. Half unconsciously we wonder how so strange, arbitrary, and seemingly random a scattering of points of light could ever have come into existence. What cosmic Mind or Law created such apparent disorder? But astronomy has shown that the stars are not arranged arbitrarily; there are galaxies of stars, some spiral in form, and these are grouped in clusters, and these clusters perhaps in superclusters. The cosmos is not a chaos; it is a great system of systems, and "from the very beginning" some ordering principle must have been at work. Here we must be humble and confess to ourselves that when we reach ultimate questions of order we are not only very ignorant but hardly know where to begin. When did the universe begin its history, and whence came its ordering principle? Is the history of the universe the vanishing of a disorder which was arbitrary, and the emergence of less arbitrary differentiated forms? We do not know, but it looks like it.

This chapter is not a study of the history of ideas, tracing all the conceptions of order and disorder that have occupied the human mind since earliest days. All that is necessary here is to take from the physics of today and tomorrow two primary ideas: those of spatial disordering and ordering tendencies, that is, processes moving toward a terminus of maximum disorder or of order. For these are not only the most interesting but also the most theoretically fundamental of all principles concerning space and time that are conceivable today. The status of order and disorder, considered as the termini of tendencies, is today the supreme issue overriding all others in human, philosophical, and scientific importance. Of that I hope to convince you.

To an open mind, unprejudiced by past or recent theories, there appear to exist in the material universe two great, and *apparently* opposed, general tendencies,[2] of which it is natural to regard all other processes as special cases:

A. TOWARD DYNAMICAL DISORDER called *Entropic*. This tendency is an abstract theoretical construct in what is called "phase space."

B. TOWARD SPATIAL ORDER called *Morphic*. This tendency is immediately observed in (3D, S).

These two tendencies have certain features in common. For example, their mathematical representations must be global variables, that is, variables associated with the totality of a system, and they are more comprehensive than laws of conservation to which they can reduce in special cases. There is also a significant contrast between them: while "A" leads only toward a statistical equilibrium subject to fluctuations, "B" leads to an equilibrium structure which can serve as a stable basis for later processes. "A" peters out; "B" leads on.

"Entropy"[3] (Clausius, 1865) is from the Greek, meaning "generating a transformation," and the entropic tendency toward greater dynamical disorder of atoms, molecules, and so on, in an abstract "phase space," was discovered around a century ago. It represents a tendency of thermal systems, starting sufficiently near equilibrium, to move toward an equilibrium state of maximum dynamical disorder and uniformity (subject to fluctuations, some of which create transitory regions of order!). The entropy tendency, when powerful enough, disrupts otherwise stable units. Believe it or not, for a hundred years this has been the only one-way tendency recognized by the orthodoxy of physics and represented by known physical laws. The result was that many imagined that it was the only one way tendency in the physical world. Actually it was the only such tendency that had been systematically studied as a class of processes with many special cases. The "opposite" process had been relatively neglected until recently, because it usually occurs in open systems.

"Morphic"[4] (Whyte, 1966) is also from the Greek, meaning "generating order, form, or symmetry in ordinary 3D space." We must stop saying "God geometrizes," which is doubly misleading, and replace it by "nature is morphic." Special cases of this class of processes have been known for long, but only now has this class been considered worthy of a name! If, as cosmologists find it best to assume: "In the beginning was Chaos," more precisely a disordered primeval gas or a fireball, then every identifiable object which human eyes have ever seen was formed at some time by some type of morphic process. All visible forms, from crystals and organisms to spiral galaxies, and all human artifacts were formed by morphic processes at some level in the hierarchies of nature. "Morphic" is a little word of great power. As Shelley said, "A single word may be a spark of inextinguishable thought." (Terminology is very important; only one thing is even more important: the selection of variables.)

The universe confronts us with this obvious but far reaching fact. It is not a mere confusion, but is arranged in units which attract our attention, larger and smaller units in a series of discrete "levels," which for precision we call a hierarchy of wholes and parts. The first fact about the natural universe is its organization as a system of systems from larger to smaller, and so also is every individual organism! This is the first thing God would have had to make up his mind about if he had created the cosmos, though it is the latest man has taken seriously.

The principle that well formed terminal states can arise from less formed initial ones has long been recognized and many names have been given to it in different contexts at different times. The most influential have been *δύναμις διαπλαστική* (formative faculty), *facultas formatrix* (the same); morphogenesis (organic), formative actions (general, philosophical), and negentropy (Schrodinger's name for structure forming processes, wrongly restricted by him to the organic realm). But none of these terms has the general and precise meaning expressed by"morphic," which covers all processes in which single less ordered systems are unified into one well ordered unit in 3D.

To give the term "morphic" more significance, choose for yourself examples that appeal to you. All formation is morphic. The formation of a crystal nucleus and the growth of a crystal. The formation from water vapor of a raindrop or a well shaped cloud. Of the globular earth and nearly planar solar system from some less formed earlier system of physical units of some kind. Of a new structure by the self replication of a chromosome. Of flowers from a seed, or of a baby from a fertilized ovum. Every finite thing that man can observe ipso facto has form, and was, we find that we must assume, formed sometime from something less formed, and the term "morphic" is no more than a convenient term to cover all these particular cases and countless others. This term is today a necessity for rational thought about nature.

Believe it or not, the great science of the West has only now given a name to what can legitimately be called the most important class of natural process, without which there could exist no form, no life, and, I would add, no mind! We should be humble. Incredible as it may appear now, and amazing as it will be to future historians of science, the most common type of process in the natural universe, and perhaps also in our brains, and the most important for us now in the 1970s, has only just been given a name.

The name "energy" was once vague and tentative but proved highly fertile. Similarly one cannot think about form generating processes in general until they are given a convenient name that provokes our attention and opens up new problems for study. *This little word "morphic" contains a universe of meaning which a generation will not exhaust.* As we shall see, morphic concepts may prove more powerful than those based on conservation or invariance. It is a world within a word. It unites formal knowledge and human meaning. Modesty or caution would here be irrelevant, for, I repeat, the idea is not mine, but universal. I chose the term, but under compulsion to find one.

"Oh, time, strength, cash, and patience," as Melville cried, so that I could unfold all the richness of the morphic way of thinking! Here is one point, to me grand. The term "morphic" calls our attention for the first time as a general issue, without restriction to special cases, to a great and ceaseless drama in nature as now conceived: *the apparent contest of two tendencies, toward order and toward disorder.* We see the universe as the arena of a contest between these two great antagonists—at least so it seems on one level of analysis. Moreover, these antagonists, though seemingly opposed, are in certain respects so closely related that in some sense they *seem to* belong to each other. There is certainly a deep correlation *between them*. Neither can be eliminated without changing everything.

The complementarity of opposites and the conception of an indispensable antagonist is an ancient theme. In Isaiah 50:8, we find: "Who is mine adversary? Let him come near to me." Skipping Plato, Aristotle, and the centuries, there is Burke, writing in 1790 on the French Revolution: "He that wrestles with me strengthens our nerves and sharpens our skills. Our antagonist is our helper." The scientific intellect can draw hints from any quarter, and we are thus led to ask: In the most fundamental analysis are the entropic and morphic tendencies really exact opposites?

Here we are well over the frontier of established knowledge, at a point in no man's land to which no right of way leads (as yet). But several arguments that the layman can understand suggest that the two tendencies are not opposites; that, at a fundamental level, one may prove to be more general and to contain the other as a special case, in spite of their apparently opposite character when expressed in vaguely defined terms, such as order and disorder:

1. The terms "order" and "disorder" reflect a state of knowledge; in a deeper mathematical analysis the opposition might vanish, as the above definitions suggest.
2. Fundamental physical theory has continually advanced by seeking unified laws covering apparently diverse phenomena. Cannot one of the tendencies be seen, in a precise sense, to be the more general of the two and to contain the other as a special case?
3. The morphic tendency is geometrical, in 3D, while the entropy tendency is dynamical, involving mass. They are therefore not exact opposites, and the morphic is more general.

4. The increase of entropy can, in many standard examples, be represented as the *decrease of a quantity*, such as a difference of pressure or of temperature, while the morphic tendency can be expressed as a *decrease of an asymmetry*.[5] Thus the two tendencies are from one point of view very similar. The entropy tendency generates uniformity (subject to fluctuations) and the morphic tendency symmetry. The entropy tendency expresses the equipartition of energy, the morphic the equipartition of space (in symmetrical forms).

These points suggest that it may turn out that the entropy tendency can be treated as a case of the morphic tendency, *but in a special class of systems*. If so the great contest is not between two opposite principles, but between *contrasted coexisting systems*. It may be the coexistence of contrasted systems, not the clash of opposed principles, which accounts for much of the complexity of the universe. There exist countless systems, and neighbors often tend to disturb one another.

This is a thought whose power will not quickly be exhausted. A single *monistic law can generate conflicts in contrasted coexisting systems*. There can exist a single universal law and yet widespread conflict. Some have sought for a quasi satanic principle in nature, generating all that is ugly, destructive, and hateful to man. This is unnecessary. The mere fact of the coexistence of contrasted systems can generate disharmony. This is one of the deeper ironies: *togetherness often implies clash*. But back to our theme of order.

The existence of systems marked by order, symmetry, or form is from one point of view the "first" fact about the universe, its most general property. This is obvious enough, though the class of processes which generate them has still to be systematically studied. But the "second" fact about the universe, which is really an aspect of the first, has become obvious only in this century, and it has important consequences.

As we have seen, these well formed systems are ordered in hierarchies,[6] in the great single hierarchy of the inorganic realm and in the myriad little hierarchies in every organism, which two types are very different. All I can do here is to list some of the units in these hierarchies as they appear prima facie, prior to detailed analysis. The known universe may be arranged in superclusters, or clusters of galaxies, and these in suns or solar systems, and (in our case, the earth) in geological formations, minerals, crystals, molecules, atoms, nucleons, and whatever ultimate particles may one day be found.

That is a quick view of the single inorganic hierarchy of levels of structure from large units to smaller ones. In sharp contrast to this unique hierarchy there is the internal hierarchical structure of every organism, from the organism itself to organs, tissues, cells, micelles, biomolecules, and atoms. This summary statement may not be complete, but it causes one to think. *There seems to be one very general type of ordering nearly everywhere in the universe!*

The contrasted laws of these two types of hierarchies relating the properties at each level are still unknown, but this "second" fact about the universe is beyond question. The universe is highly structured, both as a whole and in each organism, in a series of levels, each of which is marked by a unit of characteristic form which must have been generated by a morphic process at some time in the past. But we know that the universe of stars and galaxies is of immense age, perhaps nearly as old as the universe itself. We also know that there have been organisms on this earth for almost as long. How is this possible if—as we have been taught—a universal entropic tendency has been always and everywhere tending to disrupt ordered systems and to establish disorder? The fact which we cannot, it seems, deny is that *over vast regions of space and immense periods of time* (at least since living systems began to emerge on the earth) *the tendency toward disorder has not been powerful enough to arrest the formation of the great inorganic hierarchy and the myriad organic ones.* The conditions of the universe and of this earth have been on the whole favorable to the morphic processes.

By the way in which it was expressed the entropic tendency toward disorder hypnotized physicists for a century into thinking it fundamental or primary when it is not! A hundred years of physics has to be corrected. Entropy applies only to closed systems, which are rare, and not even to all of those! Such mistakes can happen, and they can be corrected only by looking at the immediate facts with eyes undistorted by the lenses of those theories that happen to be in fashion.

We have reached the unavoidable and significant conclusion that in the cosmos by and large the long neglected morphic tendency has predominated over the entropic tendency, and this has certainly been true on this earth since the solar system was formed. There are of course regions where, owing to special conditions such as a high temperature, there has been a tendency toward disorder at a certain level of analysis. But the ordering tendency must have prevailed on the whole, or there would exist no well formed systems, and no hierarchies of systems. I emphasize: *From a human, philosophical, and scientific point of view the ordering tendency in nature is more important than the disordering tendency.* It is for the physics of the future to discover the precise conditions—I have called them the "morphic conditions"—that determine which tendency dominates over the other in any situation.

But may this conclusion not be only a consequence of a human preference for order? Is it a truly objective inference?

We have reached the stage in the history of thought when we can no longer say anything is "merely subjective." Certainly this conclusion is not merely subjective, for our theme is at once subjective *and* objective. Neither realm can do without the morphic principle. It is represented both in the mind's search for order and in external nature's visible bias toward order. Moreover, a glance at the history of physics reinforces the conclusion that the morphic tendency does not represent merely a human bias.

The original bias of the human intellect and of the scientific mind was toward *invariance*, permanent entities, conserved quantities, and symmetries. Scientists tend to establish traditions and to form schools, and the first great historical school relied on conservation principles. So the principle of the conservation of energy was gradually established as covering many different types of energy. Energy developed from its vague beginning into a universally conserved quantity, in those systems which could be regarded as "closed." This success so flattered the scientific mind that it came to imagine that closed systems were the only important ones so far as general laws were concerned! Nature, it was thought, had given them a prior status, but really man had done this. This conviction grew into a dogma so universal as to be Unquestioned: the fundamental laws treat of closed systems. The argument for this was partly circular, but that did not matter, for the principle was fertile and led to the empirically confirmed rules for the transformation of energy from one kind into another.

This trick worked well for many decades, though Einstein showed that it was really mass energy that was conserved in closed systems. But now comes the shattering fact: *When the highest accuracy is used it is found that closed systems are exceedingly rare!* Nature does not seem to care much about them. Nearly everywhere systems at finite temperatures are losing energy by radiating outward; it is very difficult to hold a system at a strictly constant temperature; there is nearly always a leakage somewhere; our sun and probably all the galaxies of suns are perpetually pouring the energy of radiation outward into space. Or so it seems.

Here the difficulty arises that such considerations are affected by the kind of general theory of the cosmos we choose: stationary, expanding, or cyclic; with mass being created, or not, and so on. But there is a powerful argument which suggests that open systems losing energy (or displaying decreasing asymmetry) should receive more attention. To show this I shall make use of two technical but easily understood logical terms, from the theory of relations.

An "asymmetrical (two term) relation"[7] is one incompatible with its converse. If I am older than you, you cannot be older than me. On the other hand, a "symmetrical relation" implies its converse. If I am the same weight as you, you must be the same weight as me. Now *asymmetrical relations are more powerful than the corresponding symmetrical relations which they contain as limit cases.* If I am heavier than you, but only by some negligibly small amount, then I am, in effect, the same weight as you. Where the asymmetry is negligible, the asymmetrical relation becomes a symmetrical one.

Now take a deep breath, for we are about to enter a new realm. If the laws of nature are best expressed in asymmetrical relations, that is, an asymmetry at t_2 is less than it was at t_1 or $A_{t2} < A_{t1}$, it is in general easy to treat principles of invariance as limit cases of some principle of one way process. Indeed laws expressing general one way tendencies in nature are necessarily more powerful than the traditional conservation or invariance laws. We have expanded our territory; our new realm contains the old one, and more.

In logic, that is clear. But it is difficult to demonstrate it in particular cases. This is because ever since the ancient Greeks, and particularly since 1600, the Western intellect and later exact science have cultivated the intellectual craft of thinking about invariance, permanence, symmetry—many kinds of symmetrical relations. So it is very hard indeed to change such deep habits and to take asymmetrical relations and one way tendencies seriously. In fact entropy is still after a century to many an obscure and troublesome idea, and it will remain so until we have had more practice using asymmetrical relations. Entropy troubles us because our Western minds have practiced placing the emphasis on symmetrical relations for some two thousand years. Equations seem natural; inequalities seem awkward. We cannot reason properly using asymmetrical relations until, *first,* we are absolutely forced to do so by the patent inadequacy of symmetrical relations; and, *second,* some lucid model shows us how to reason using asymmetrical relations and how quantitative and symmetrical properties can be derived as limit cases from ordering asymmetrical relations. One specific example in fundamental theory is necessary, but unfortunately this has not yet been achieved by anyone. So far we have not reached the threshold from which the going downhill is easy.

Notice that this means that scientific thought, thus far, has *not* been biased by any desire to discover a tendency toward order, but by a preference for invariance and conservation. Moreover, the first tendency discovered was the entropy tendency toward disorder, clearly not the expression of any preference for order. Viewed historically, it is thus a point in favor of a tendency toward order being objectively valid that it is now being proposed in spite of the historical factors working powerfully against it. Indeed a tendency toward order is only now acceptable because at long last it provides a universal principle prevailing everywhere (where entropy does not override it), by which I mean in the subjective and the objective realms. We have stumbled blindly onto the clue uniting these two: a tendency toward order!

But so radical a revision of methods has many significant consequences. *First,* order and disorder are treated as the *immediately given relations* of (3D, S). Here we follow Leibniz, who wrote: "*Space* is nothing but the order of things possible at the same time, while *time is* the order of existence of things possible successively." *Second,* order and disorder are *global* terms, applying to the totality of the system, not to any one of its parts. To risk terms which may mislead, it is of a holistic or gestalt, rather than of an atomistic, character. *Third,* the two tendencies imply a movement toward internal equilibrium and are thus *finalistic,* though they do not in general imply a purpose, an aim, a final cause, or any vitalistic elan restricted to the organic realm. There is a *nisus* toward a terminus, but wider inferences from this may be misleading: *Fourth,* the terminus is of great importance and interest, and introduces a *valued qualitative aspect.* If the terminus is one of higher order, we value this property; the presence of a tendency toward order is necessarily valued by man. We like crystals and shapely organisms.

From the earliest days order has been regarded as a supreme value. God is order; he alone creates and restores order. The supreme Order is divine. It is inherent in human nature, since man is an organism, to value order. But this is ambiguous; the experience of order can mean different things at different levels of the mind. Since we lack a clear conception of the various mental levels, I shall here discriminate only two modes of experiencing order. There is, first, the experiencing by the total human person of the presence of an ordering tendency in *which he himself participates.* This is a powerful emotional experience: awareness of the power of the ordering principle within and without.[8] But at a more superficial level there is the detached intellectual and aesthetic recognition of some established form of order outside ourselves. This is the perception of order which has until now provided the basis of science. But these two–the emotional participation in the great ordering tendency within and without, and the supposedly detached and objective observation of order in nature–are not separable in the integrated individual, for in him the various levels of the mind are in continual interplay.

A final question we may ask is: Are these ideas about the ordering tendency not only reasonable, but timely? It seems that they must be. At some periods in the history of science advance suddenly becomes rapid. The idea of hierarchical structure in the universe and in organisms has been obvious to some for many decades; during the 1960s it suddenly became topical to frontier minds; in the 1970s it is already commonplace as a new community insight. The next step is to ask how hierarchies came into existence and prevailed so long. Thus it is hard to imagine that the concept of a hierarchy of morphic processes will not in turn be commonplace in a few years. The great task of the coming decade for physics and biophysics, and perhaps for embryology and for understanding the development of the thinking brain and of language, is *to discover the hierarchy of morphic rules relat*ing *all levels in all relevant processes.* The 1960s discovered hierarchy in the universe; the 1970s must follow on by discovering the morphic processes which have created these units at all levels and shaped the hierarchies.

In this chapter I have emphasized the dominance of the morphic tendency, but in certain circumstances and at certain levels entropy wins; tomorrow a hundred bombs may disperse most earthly organisms back into their component parts. Moreover, there is no perfect ordering, no perfection of structure in the natural universe if we exclude isolation near absolute zero.

Yet for this assertion to be fully valid, we must exclude consciousness, for human beings can be aware of a subjective perfection. This is a valid experience, though of a different order. The conscious mind can experience for moments, perhaps for weeks, I would say never for a lifetime, an inner perfection. This is a fact as much beyond question as any other in this universe of experience. A science which neglects it is incomplete. Can we construct a view of the universe which is complete, at least for us in this century?

4

The World View

A view of the universe of ordering processes goes deeper than science or religion and can provide both a nonsectarian ideology and a clue to unity in many realms

"The world has need of a philosophy or religion which will promote life."
—Bertrand Russell [1]

"I am convinced that there is only ONE *basic* Order, *which appears as logical or mathematical to our cognitive intuition, aesthetic to our emotional intuition, and moral to the volitional or conative.* AND *it is essentially 'numinous.'"—Cyril Burt* [2]

Now I approach the core of my argument, clarifying what has already been suggested. The aim is to use the most general features of the known universe to enable us the better to order our thinking. Wherein is nature better ordered than our own minds are today, prejudiced as they are by faulty traditions? What pattern of ideas, or basic world view, can we derive from the universe and thereby understand external nature and our own minds better? What is primary in external nature may also be primary in human thought. If so, we can make a step toward the unification of nature and thought. The ability of mind to understand nature will then no longer be a puzzle.

The Eastern contemplative emphasizes unity but scorns detail, whereas the Anglo Saxon empiricist dislikes generalizations which seem to go beyond the facts and is suspicious of ideas. We must pass beyond these prejudices and seek a way of thinking which gives equal status to general law and to particular facts. Equal emphasis must be given to the properties of wholes and of parts, and also to subject and object. We require an elegant view of the whole which does not lose the details, overall principles which permit a myriad applications and give us balanced understanding.

There is throughout the world a hunger for timely ideas, hidden and unexpressed in China and Russia, half conscious in Europe, fully aware of itself in the United States, and pathological in California. But everywhere, at some level, there is a sense that there ought to exist ideas which can bring relief to the individual and new power to emotion, thought, and action. It is as though mankind, near despair, felt that God, or nature, or history, had failed it and left it today, in this terrible century, lacking what it should possess and enjoy. If we are thinking organisms, should we not be able to think better?

What follows is in no sense mine; it is universal. Here my skepticism and irony vanish, and I make high claims. I offer a world view which, once fully understood, can unite conviction, idea, and action in those young in spirit whom it seizes. It is both a philosophy and a way of life, and it has to be understood through its applications. As there is no longer a clear separation of philosophy and science, I prefer to regard it as philosophical. It is not new, but the emphasis given to it here is without precedent. The reader will understand that some repetition of the principal melodies is appropriate. This work does not pretend to be a systematic formulation for theorists, but a joy and relief for men and women and a foundation for science.

The world view is more than a skeleton of a theory of the universe, for it is a view of the relation of the human heart, mind, and will to universal factors. It is the germ of a new psychology, one which sees the human mind to be as much a part of the universe as is any other part. If an individual can read this book with a favorable imaginative response, as the vista of a possibility, he is already on the way to becoming an individual in a new culture. To think, to feel, and to live as the world view implies is appropriate involves a new discipline which I will not prejudice by calling it "spiritual," since the world view transcends such categories. When I describe the world view as "beyond science and religion," I mean as these two aspects of culture have mainly been understood in our century. There have always been wiser spirits who saw deeper. The view covers a fertile principle which we cannot, without lessening its significance, call "spiritual," "organic," "scientific," or "psychological." For it expresses a philosophical truth about the universe of experience which, by absorbing man, transcends all past categories of systematic thought. I believe that no previous philosophy has identified a single principle operating in external nature and in the human mind.

The world view, though it absorbs what is valid in human experience of perfection and unity, is not a pantheism. For this experience does not pervade the universe. It arises only where and when a member of our species is in the appropriate condition.

All thought and action rest on prior and often unconscious metaphysical assumptions. Here is a philosophical view which is fertile because it recognizes and so facilitates the creative interplay of many levels of the mind in the working of the imagination. It eases the reunion of "conscious" and "unconscious" which has been the deepest aspiration of the West in our time. It points the way toward overcoming the separation of subject and object by identifying a principle common to both.

But its emphasis is new, and it will not easily seize tired minds. It is the whole natured intensity with which the individual identifies himself with the world view that renders it fertile. Accepted, it becomes a joyful instrument which can bring intellectual clarity and enhance life. But I can communicate nothing which is not already latent in your thinking, conscious or unconscious. The intensity of my conviction does not guarantee its truth, but convictions are infectious, and mine serves a necessary purpose, in me and in you.

Some hold that deep in the human psyche lies a core of rebellion out to destroy man and civilization. At the taproot that cannot be true. Self-love in the strict and narrow sense cannot be the root of human nature viewed as the deepest potentiality in human heredity and in the human mind. Since we are organisms, beneath all psychoses; neuroses, perversions, and egotisms there must lie a morphic principle promoting coordination and order. The world view identifies this deepest potentiality of the human being and his mind with the most general feature of the external universe, and—as we shall see—by doing so closes the circle of knowledge. This circular closure is possible, because while mind is a part of nature, nature is also an image in the mind.

We now come closer to the world view: Its aim is no less than to reestablish reliable principles in a time of rapid change, relativism, and uncertainty. Man has never before known so much about himself as he does today, but what he knows concerns mainly particular differentiated faculties and his contemporary pathology. The world view presents to man a view of his healthy global capacities which he has never had before. If the world view appeals to any individual, he can accept it as a basic conviction providing both the foundations for a culture and a heuristic principle helping him to discover what is true for himself as a unique individual. It is a free gift from the universal.

Moreover, the world view potentially covers valued qualities as well as measured quantities. Romantic minds have sought a human meaning in the universe and have tried to regard it as favorable to man. The world view shows in what sense this is true; and in what sense false or meaningless. In it man for the first time acquires rational principles concerning external nature whose structure corresponds also to the structure of the operations of his own brain. Thus, at last, he can in a clear but restricted sense feel at home in this universe, while knowing that any moment may be his last and scorning the spurious consolations of transcendental religion. Moreover, the world view provides what has long been sought: a rational foundation for man's deepest personal convictions.

Some patience is necessary in considering this new vista; everything cannot be seen at once; many difficulties in current thought have to be reexamined on a new basis; a new language and a new morphological way of reasoning have to be learned. But some features are simple and clear, and can be seized intuitively. The world view asserts that the known universe as a whole, and every organism, including man, contains a graded sequence of units in each of which a formative tendency has been, or still is, present. Nature is everywhere creating forms when conditions permit, just as there is an order-generating tendency in our own minds, when not pathological, this mental tendency being a particular expression of a universal tendency. *Natura naturans* is a workshop of forms, and the world view of nature not only a philosophy but the basis for many future sciences of form in inorganic, organic, and mental nature. Allow these first hints to sink in. This book, read with comprehension, should bring some immediate benefits of a movement toward clarity, consistency, and unity. But its myriad applications will take decades to unfold.

We are here, I believe, at the central principle which in man not only underlies the capacity to generate languages and to create philosophy, religion, art, end science, but enables these aspects of culture to be seen as only partly separable elements of what is fundamentally, or can be made into, a single unified tradition. Moreover this is not the arbitrary dream of a maverick in love with unity, but a timely working hypothesis, the fertility of which will be tested during the lifetime of my younger readers. If my judgment is correct, no overreactions by overspecialists matter at all; they are inevitable. But notice that I do not, like the founders of some religions, claim transcendental authority. I simply assert that this view represents today the optimal path toward the unity of knowledge, as one aspect of the urgently needed integration of man as individual and community, and that the effective dominance of morphic processes throughout the universe will be recognized by all competent young minds during the present decade. The universal forges what is timely.

The search for a single comprehensive order, whether made consciously or not, is as old as human thought. Here is one man's expression of a universal idea of which he is the carrier. Everything that is new in this book, or new in its emphasis, arises from one fact: as the carrier of the universal I have been compelled to take fully seriously, perhaps more so than anyone else recently, the philosophical and scientific consequences of one unassailable conviction: that a radical unification is close at hand affecting both knowledge as a whole and each science individually. This conviction may or may not be valid or timely. That cannot be rationally known until tomorrow. In the meantime I am guided by many indications which appear to me to justify my convictions. As I have said, my guardian principles are unity, clarity, and immediacy. To achieve clarity and unity we must return to the immediately given three dimensionality of space, and the experience of temporal succession.

The principles of the world view are implicit or explicit on every page and chapter of this book. But we have now reached the point where a definitive formulation is appropriate.

The world view sees the structure of the universe of experience as a hierarchy of morphic processes, i.e., as a sequence of levels of formative processes in 3D and S, evident both an external nature and in our minds. To understand anything in nature, without or within, at least two levels must be considered. In this world view lies the clue to the unification of knowledge, of the various sciences, o f the human person, and possibly, for this is only a personal hope, of the human community—all this in this century.

At a unique moment in intellectual history one must not hesi-tate to speak without equivocation. If there is an authentic unity in nature accessible to us today, the world view points to it. In a given epoch a synoptic view of this comprehensiveness can come into sight only once. Our supreme need today in the 1970s is for a lucid synthesis of experience which can, at least potentially, clarify human thought and action to express man's deepest desires.

This world view is certainly not yet axiomatic. In it we step back from the specialist abstractions, however valid they may be under certain conditions, and set out to construct a lucid and unified approach to what is immediately given in 3D and S. Here I cannot analyze all the different levels on which the morphic tendency is evident today, or will be soon; that would carry us too far both from my own competence and from the main purpose of this book. Nor have I examined current theories of cosmology, because I believe that all theories of the history of the cosmos, of the galaxies, and of suns are, for one overriding reason, provisional. All such far reaching theories in space and time must require radical reinterpretation when a unified theory of particles and forces is available. What I offer here is a general method of approach, perhaps more comprehensive than the Organon of Aristotle or the Method of Descartes, as it includes man himself. It is a manifesto for the late twentieth century.

This view, or something like it, has been expressed before,[3] but not in this context or with the same emphasis. The fertility of the world view does not lie in its novelty, but in the primary status given to it. Through it and through it alone can the human mind advance in the immediate future from current confusions toward a redeeming clarity. If it is valid and timely, current scientific theories lacking unity will, I assert, be replaced during the coming decades by a philosophical unification of knowledge as a whole, and a scientific unification of each science separately,[4] based on this world view. This implies a radical metamorphosis of science, guided by the criteria of unity and immediacy. This in turn implies, as I have often already explained elsewhere; the development of a science of developing forms, leading to a unification of physical theory, of biophysical theory, of general biology, and of psychological theory, based on the concept of morphic or ordering processes. In such a family of unified theories of developing forms, the mechanical concepts of the "actions" and "interaction" of one localized material entity on another, and of their "relative motions," are replaced by one master concept: the relaxation of extended spatial forms toward symmetry, the last traces both of mechanism and of redundant abstractions being discarded.

No more severe test can be applied to an emerging philosophical and scientific theory than to require of it such achievements within decades. If the world view is valid and timely we are on the eve of a torrent of theoretical discoveries facilitated by this heuristic principle: the universe of experience is to be regarded as a hierarchy of levels of morphic processes.

However, a personal word of warning to the ecstatic, and of assurance to the timid. This world view will accomplish much, but it will not eliminate agony and tragedy from any personal life. It will simply help man to make the best of things, by eliminating many unnecessary lesions and confusions and much ignorance. If it were my last word, I would say this: The morphic view is the most powerful redeeming factor conceivable to any human mind in this century, an authentic promise of real possibilities, provided it is accepted with. patience and with readiness for frustration. The human mind is built on inertia and does not turn a major corner easily, but the world view brings this blessing: it is not an intellectual system imposed by a human mind on external nature. It is nature and mind showing us their common form and thereby washing away many past mistakes. I would not use such language if I did not believe that I know that history would justify it, and if I had not already learned to live with this knowledge. But back to our theme.

The universal morphic process generates the coordinating tendency of organisms and the order seeking, tendency in the human mind, and in all of these the morphic tendency operates on levels forming a hierarchy. But this is no universal frictionless process. For in certain systems the tendency of the whole works against the tendencies in its parts, and so conflict and disorder can emerge. In the next chapter we shall examine how the apparent conflict of these two tendencies can become the constructive interplay which generates and maintains life. But here we are primarily concerned with situations where this conflict and interplay do not arise.

No single human mind can be an adequate interpreter of so grand a theme, unless he is guided by a deep necessity. It is usually thought that it lies beyond human power to anticipate the future of science. At certain moments this restriction vanishes, and an authentic vista brings the future within sight of the present. The Greek atomists saw far ahead. Moreover, for a century or more the guiding rule of the exact sciences has been: Search for invariants! From now on, I suggest, the supreme heuristic rule for scientific theory will be: Identify the inorganic processes at each and every level, everywhere, outside and inside man. In this specific sense I am foreseeing the future of the sciences. This transformation of the dominant aim of inquiry marks the present decade as a turning point in the history of the intellect, for many other minds are working in the same direction. We are all part of one tradition, now in rapid transformation. I regard the world view as a collective insight, though I may have had a special role to play.

As we have already seen, the world view is more than scientific, in any strict traditional sense. For it concerns the entire universe of experience, subjective and objective; it is postreligious, and prescientific; and it treats order and disorder in so general a manner as to be prior to all testable scientific principles. We shall see that it achieves this through a circle of applications, in contrast to the linear deductive form of recent scientific theories.

For some the world view, as being concerned with order and disorder, may appear to be merely a superphysics seeking to absorb everything. It would be more correct to call it a metaphysics, departing from traditional physics so profoundly as to change its character and to require it to be called a philosophy. I emphasize again that in the following respects it contrasts with the physics of the past:

1. *The world view as finalistic*, in the formal sense of being concerned with tendencies toward a terminus. This does not imply necessarily, or in general, what has been understood by a teleological process involving a final cause; but it does take a step away from past physics toward the needs of biology and psychology. In doing so it salutes Aristotle.
2. *It treats global properties* of order and disorder, associated with the totality of a system, rather than local atomic properties, such as mass or electric charge. This again is a step toward global properties of living systems such as the life cycle, differentiatedglobal functions, and gestalt and similar properties in psychology. But it raises the basic question of the relation of the world view to the success of the analytical method, which cannot be discussed here.
3. *It is relational*[5] rather than *substance bound*; it deals with changing patterns of relations in 3D, rather than motions of masses. *Changing 3D forms replace material motions*. This suggests that all branches of knowledge will be illuminated and simplified by a new metaphysics of relations, the scope, power, and implications of which can hardly be imagined today.
4. It displays the *immediacy* of 3D and S, in place of the abstractness of the basic concepts of contemporary physics:
5. Finally, and this distinguishes it essentially from much recent philosophy and science, it discards the aim of hypothetico-deductive theory passing from arbitrary postulates to their necessary conse-

> quences, and asserts that today the aim of all intellectual endeavor should be the development of a *closed circle of* concepts,[6] self consistent, comprehending all twentieth century knowledge; and without arbitrary features. Reasoning. can start anywhere in the circle, say with man, who observes the universe and identifies universal laws (as yet unknown), which in certain circumstances (by a process not yet fully understood) permit the emergence and evolution of organisms, and then (also by a still mysterious process) the emergence of a language using thinking species man, and the circle is closed. In this circle of knowledge man has a unique status. He closes the circle, for the organic laws permit his evolution and it is he who experiences himself and observes the universe. This circle is possible and can be closed, only because a single morphic tendency plays a crucial role in determining the advancing steps all the way round the circle, all segments of the circle being systems of relations.

The world view is thus circular, relational, global, finalistic, morphic, and possesses immediacy. The objectivity of real systems "out there in space" no longer requires a basis in "matter"; there exist only changing patterns of extended spatial relations, our awareness being of these. The completion of this relational and circular world view, with its offshoots the various sciences, should be the supreme all encompassing aim of the intellect at this stage in its history.

Like all basic advances, this view challenges established orthodoxies, demanding a painful stretching of the mind. Every radical unification requires compromises from several sides. This one demands a morphological awareness and way of thinking, transcending the overanalytical, overstatistical emphasis of much current physics and biophysics. Moreover, it implies a surrender of independence from all the special sciences, which they will hardly accept in advance of some unmistakable authentic achievement proving its desirability, and this is not yet available. It also means, and this is a lot to expect, a restoration of the prestige of comprehensive philosophy, though in a new mode. The path of advance is certainly arduous. Yet the potential gains are immense to philosophy, psychology, biology, and man, apart from those to physics, which are my main interest. A grand intellectual battle is in preparation. Let us hope that it will be fought with the composure and dignity worthy of high philosophy and authentic science.

This convergent movement toward a philosophical unification will not be achieved by continued logical analysis, but by a single major step forward, when it becomes possible, from the foundations of the separate sciences toward a shared basis. This means that the *circular self consistency* of the emerging unification becomes the supreme criterion, complementary to the quantitative or logical *precision* of the separate sciences.

Further, as already suggested, this unification must involve the elimination of the paradoxical abstractions of current physical theory, and an emphasis throughout on the immediacy of observed and experienced relations and of objective and subjective facts. The unification will be expressed in terms comprehensible to nonspecialists, because they deal with order and disorder in (3D, S). Spatial relations are primary, and in the course of a morphic process systems relax toward symmetry. But "time" as an entity or an *extended quantity* is discarded not only as often misleading but as scientifically redundant. *Temporal succession* is the observed and necessary relation behind the treacherous concept of "time."

Nor is that all. The immediacy of the world view, and its reliance on a tendency toward order, implies that this view possesses a significant meaning for man, in contrast to the formal abstractions, empty of all immediate significance and ambiguous in status, of recent theories, particularly in physics. Moreover, since order possesses value for man, there is at the source of all the sciences which arise from the world view an intrinsic value and the germ of an ethic. We escape the recent exaggerated, indeed pathological emphasis on methodology, or the mere logic of technique, and establish significance and value in the foundations of knowledge.

This is a first glance into the new world implicit in a new word: morphic. It is more even than a new metaphysics, for man is an active element in it, and this fact reaches far. Morphic man is man remade. For the world view presents no less than a transformed equivalent of a fusion of the unifying powers of the poetic imagination (including their highest expression in joy and mystical experience) with the powers of reason. I say transformed" because in the new view such terms as "poetic," "mystical," and "rational" not only acquire deeper meanings and new correlations but lose misleading associations There is, as already emphasized, no basic separation of emotional conviction, intellectual idea, and effective action—these being, in the healthy coordinated man, aspects of a single tendency, which is more than an élan vital, because it is universal and not merely organic. The aim is to leave behind the spurious elements in traditional religion and the mistaken features in past science, and to acquire a new dimension of experience richer than either. Moreover, in doing so, we put man once again at the focus of his own universe of experience, not at some hypothetical geometrical or gravitational center of the known cosmos, but at the summit of all known morphic processes. The human imagination is incomparable, no other animal possesses it, and all that man possesses comes to him from it. It is more than divine.

Let this also be clear. The world view is not a "scientific religion," both terms being irrelevant and misleading, for it goes beneath both to their common source in the human morphic imagination. The arrogant and fateful "independence of science" which has nearly destroyed us vanishes for good, as do the sentimental, parochial, and unworthy elements of the pseudoreligions which relied on rewards in heaven (God must have been deeply ashamed of his transcendentalizing prophets, who bought their following with spurious promises. Spinoza was more honest and his religion nobler: "Love God, but do not expect him to love you in return.")

It is the destiny of science and religion now to undergo a purging of the spurious, the incredible, and the contradictory, so that man can breathe freely again. Nothing less than the present confusion of basic physics, the frustration of biological theory, and more important, the agony of man on the precipice of social violence can shock the weary adult into partaking in this unprecedentedly radical self purging and rethinking, guided by a younger generation. My optimism is based on more than a mere hope, for it is my belief that everyone who rejects the spurious comforts and flattering nonsense and self deceptions involved in traditional doctrines must at some level in his nature welcome this liberating world view in spite of its strangeness and say, "Yes, it is worth a try. It cannot be more mistaken than the doctrines of the past. I must think it through for myself, be with it, and let it be with me." I have given it this format as reading for everyone, rather than as an academic paper offering a mathematical symbolism concealing a secret wisdom for specialists. The blemishes of my presentation can reassure the reader; this is clearly no exercise in abstract logic by yet another brilliant analyst.

There is, I assert, a historically unprecedented magic in this world view. Indeed it could not be otherwise. What is it that you most desire? The therapy of a loved one? The view points directly to the presence of a hierarchy of health restoring tendencies in every person, though never guaranteeing recovery. Or is it the redemption of mankind, in a universal consensus? The view points to the only basis on which all can agree and—I personally believe—to the way in which an authentic consensus can develop. Or the unification of physics? It is my conviction that the conception of morphic processes can facilitate that end. Or is it a transformation of thought so that the properties of wholes can be identified without neglecting the parts, and thereby "life" be understood? The view holds out an earnest of that. Or is it relief from all the torturing confusions of the abstract intellect today? The view offers the prospect of a new immediacy and clarity. My task is to assert these things; it is for others with specialist competence to demonstrate them in detail.

This presentation of a timely world view is for me, its Secretary, to use Blake's word, the communication of a necessity. But others can regard it as a historical challenge and experiment. The world view is a hypothesis. which will be validated or not before this century is out. For I assert, as a necessary consequence of the world view, that the complexities and limitations of current science are due mainly to the neglect of the morphic process at the relevant levels, or in more technical language, to the neglect of asymmetrical relations. Let me make this so precise that it must be either right or wrong. On the one hand I am asserting that the unification of physical theory and the formulation of an all embracing theory of the emergence and the development of organisms covering morphogenesis from genome to the functioning of the human imagination and the evolutionary development of language will be achieved, and can only be achieved, by treating the hierarchy of morphic processes as primary. On the other hand, I am suggesting that what is missing in psychology, the formulation of a theory of the characteristics of subjective experience, of the qualities of human awareness, will also be achieved on the same basis.

It is no part of my purpose to develop here a morphology of the characteristics of states of awareness. Yet I wish to give a hint of how a theory of changing spatial relations can cover qualities of awareness. Each type of awareness can be determined by the character of a changing pattern, as in music for example. The experience of music rests on such features as stationary, repetitive, or progressive patterns; harmonious patterns or patterns of discord; simple or complex patterns, linear or hierarchical forms of harmony and contrast, increasing or decreasing tension or tempo, progress and culmination, and so on. A philosophy of the changing character of a system of spatial relations can, it seems to me, provide the basis for a philosophical theory of the characteristics of the emotional aspects of subjective experience, not only in music, but in all forms of human awareness. The world view, in other words, offers the promise of correlation at the most fundamental level of two aspects of the universe of experience: the qualities of subjective awareness and the quantities underlying exact science. Indeed, if achieved, this will be a union, rather than a correlation.

Two pseudodoctrines pervert current thought: "physicalism" and "mentalism." Physicalism is based on the conception of real, permanent, or transient spatially localized physical entities, out there in an abstraction called "space:" Mentalism, in contrast, assumes states of a reality called the consciousness of a human individual. The world view passes beyond these, not to a mere tertium quid constructed for the purpose, but to a radical relationism, accounting in certain situations for the laws of physics, biophysics, and biology, and in others for the quality of awareness. "Matter" becomes an extreme limiting case, seldom relevant, and "consciousness" a characteristic of the human organism attentive to a change in pattern of a certain character. "Consciousness," in the abstract, separated from, what it is an awareness of, is as empty and redundant as matter, in the abstract, separated from the spatial pattern it determines. The world view is perhaps nearer to mentalism than to physicalism. That cannot be helped. Existence is not designed to flatter our desire for symmetry. But no label from the preunitary age, such as mentalism, can correctly be applied to a view which goes beyond and beneath all earlier categories, and bases itself on (3D, S).

No lesser historical test than its immediate fertility in the objective realm of physics can measure the earnestness with which I approach this task of presenting the world view. Thirty years ago I formulated this test in respect of a particular basic problem of physical theory.[8] I refer to it now because during these three decades no one has yet established the results then anticipated. What was too early then may be more timely now. If the inner voice and its claim are not spurious, history will justify the theme I am presenting by meeting this test. But if some more elegant solution of basic problems is found elsewhere in time, before civilization entirely collapses, I here and now declare in advance my retraction of this work, except insofar as it be found compatible with that better solution. No one who has ever lived, no religious or philosophical genius, could rightly claim that his inner voice spoke the last word; how much less can I.

At some date in the future the world view, if it proves acceptable, will bring into being a new classicism, marked by the absolute character arising from its fusion of subject and object. Late twentieth century man may not greatly err in regarding it as definitive, as far as his own epoch is concerned. This sense of finality will be legitimate if the world view indeed defines the optimal path toward the union of feeling, thought, and action and the unity of knowledge.

On human affairs the view implies no dogmatism. It leads to no universal system of values, except insofar as individuals, guided by the world view, may discover that, in spite of the tragic divisiveness of most of the great religions, there exists one universal principle of human life which is natural today to harmonious men and women. The world view may realize the transvaluation of values anticipated a century back, for it accepts life.

Let me state again, even more sharply, what is for me the most remarkable feature of the present human situation. This is a time of moral nihilism. But this generalization conceals remarkable underlying facts. Man is a thinking animal and cannot be expected to be capable of living in a manner proper to his hereditary nature unless he knows how to think. Fully adapted thought is indispensable. But this is an age when man is scarred by partial sciences which provide no acceptable foundations for thought. "Scientific man" drifts aimlessly without foundation, root, or purpose. Physics has discarded matter, but has supplied no substitute. Biology lacks clear theory, and fully understands neither life nor its evolution, nor the supreme feature which made man: the development of language. Psychology is paralyzed by the weakness of biological theory. Religion, treated as a separate venture, has failed once and for all. Traditional philosophy has proved impotent to bring clarity and strength. Man has lost his way, as in no previous civilization, for even in the darkest ages of the past a few individuals were sustained by the hope given by some religion, however spurious we now know its promises to have been. Man, as never before, is in the dark without map or compass.

The aim of the world view, in a social context, is to give all nonspecialists some clarifying vistas, a way of thinking that they may find fertile, and to specialists a foundation that may remove difficulties. It does not directly offer a solution to political and economic problems, still less a dogmatic theory of human history, but it can strengthen the only kind of man who can possibly solve such problems. Political action and revolution by force must defeat their intentions when carried out by men and women whose psyche is filled by the ugly darkness of nihilism. For then lovers of power who lack a humane aim eventually seize power from the gentle and wise. Mere Marxist utilitarianism does not express the deepest needs of the human psyche, and every arbitrary use of power for whatever supposed material end merely brings species suicide closer. The silliest error ever made by man was to suppose that material improvement of his condition was his primary motive or need. He is a thinking animal, aspiring to unity, but needing, like every animal, *excitation*, excitement, stimulation to more intense living. (Did Freud really believe that all pleasure lies in a *lowering* of tension?) Lacking anything to live by, he turns to violence for the sake of excitement. I do not forgive the politicians, sociologists, and teachers who have failed in this critical time to see the obvious. Have they not yet understood why Hitler got as far as he did? He knew that man lives not by bread alone, but needs that greater intensity of living which can be achieved either by its authentic enhancement at all levels or by comradeship in violence.

To some who think an idea is taken when its expression in words has been hastily read, the world view that I have sketched may seem far distant from the human passion for destruction evident around us today. If so, I ask them to relax and to allow some of the myriad implications of the world view to arise gradually into their awareness. But in the last resort my case is simple. Only a new faith in man as the seeker of unity, the supreme instrument of the morphic tendencies in nature, and of the coordinating tendency in every organism, can restore man to himself by pointing to his own sense of joy, to his place in nature, in organic nature, and in universal nature. Is this madness, or sanity?

The world view will, and only can, be evaluated by the young in spirit, and even such high, generalizations as these must find the appropriate channels for their effect. The United States has the highest vitality and the least contentment with tradition, as exemplified by its searching and its violence, and a younger generation who have already proved their sense of what is now necessary—greater immediacy in living and their ability to influence their country's use of its military power. I see in the youth protests of the late 1960s in America the one social encouragement of this century, for some at least of these young men and women tried to stand for the universal dignity of man, the quality of individual experience, the enhancement of living; for immediacy, joy, and love. In these protests, beneath all contingent features, there is evident a crisis of faith and authority and a religious search for a worthy aim, coupled with hatred of the tarnished Establishment.

Those who cannot see this valid core beneath all the grubby trappings of protest and wild experimentation are beyond my help. Perhaps they have not had my experiences, or read what I have read. "Make love not war" is ancient wisdom spoken with the voice of our time, and it points to the only solution: universal love in a new specific sense, that of love clarified and strengthened by an appropriate world view and a growing consensus.

Finally, I come to what may be for many the hardest insight of all. Nothing great in the human past came into existence because some man or group believed in the "likelihood" of its "success." These terms are bogus, and the utilitarian pragmatic attitude they express is man's worst curse. Must I remind anyone that there exist no determinable probabilities in important human affairs? That is true, but trivial. The only criterion for action is its expression of man's whole nature, its beauty as a sign of coordination in the individual, and its necessity for him. I have written this book, as I wrote *The Next Development in Man* thirty years ago, because I had to, not because I saw hope of a new social "movement" tomorrow, or some visible political "success" the day after. Here not only the poets, artists, and prophets are with me, but also those ordinary men and women everywhere who do their best, though there is hell around them, because their nature tells them they must, not because "God's, in his heaven," for they are no longer that stupid.

A number of writers concerned with the long term view of the future of man have suggested that the next stage in human history will be away from the recent "materialism" toward a new "spiritualization." I confess that for me these distant dreams lack immediacy and concreteness. Important historical changes, as I see them, are brought about by highly specific factors: a personality, a challenge, or an idea, not by generalized fears or aspirations which are always with us. Though I believe the world view may possess an element of lasting truth, it is presented here primarily in relation to the desperate need of this century for a new alignment.

We must first ask, before we come to man today, can the world view assist the approach to a theory of organism?

5

Organisms

In organisms the two tendencies cooperate to generate pulsations and to develop and sustain life

It is unfortunate that the use of the above chapter heading may appear to separate organisms from the wider context, both of their past origins and of their present environment. I shall assume that a modulated historical continuity linking the inorganic and organic realms is a necessary feature of the history of life.

Since the time of the ancient Greeks, organisms have fascinated the human mind. This is not because we are ourselves organisms, but because in organisms the parts appear to be in some sense subordinate to the whole and to cooperate so as to sustain life. How does this apparent cooperation come about? The existence of organisms presents a challenge to the intellect as far reaching as any in the history of thought, and one that has not yet been met. It is surely significant that after so long, there is still, no satisfactory theory of organism.[1]

Today the problem of organism can be formulated more precisely. How can processes in a system containing untold myriads of atoms generate simple effects, such as those of the life cycle? How can systems which are far too complex for complete analysis display global properties of great simplicity? These questions represent the most advanced way, known to me from the literature of the 1960s, of formulating the problem of, organism.

What follows is more speculative; we are now moving into the front line of the 1970s. It would seem that one step toward an answer must lie in treating as theoretically fundamental the high internal ordering of organisms. It is clear from recent structural studies that the functional regions in organisms consist of very accurately and subtly arranged and coordinated parts: the atoms, molecules, polymers, micelles, and so on. Moreover, this internal ordering is to a high degree preserved when changes occur in the relative spatial arrangements of these parts in the course of growth and function. It is reasonable to expect that in a very highly, ordered system, but one that is labile, not like a rigid crystal, processes which involve the relative motions of myriads of atoms can nonetheless generate simple global effects, and there is evidence for this in recently discovered effects in highly ordered inorganic systems.[2] We are tempted to conclude that the well ordered organism can do things as a whole without disturbing the majority of the relations of the parts. *Some parts move about and the organism functions, and yet the underlying organic ordering is undisturbed.* Herein lies the problem.

That formulation may seem to be satisfactory, as far as it goes, but it hangs fire, as no one yet knows how to express it in terms of atomic geometry or biochemistry. Certainly no philosopher of organism, from Aristotle to Whitehead, has thrown any light onto this natural wonder. For me as a lifelong student of atomism, it is truly a miracle. The doctrine of atomism in its traditional form asserts that the constitution of material systems is atomic, and that their properties are reducible to changing arrangements of atoms or other small particles *possessing invariant physical properties,* such as electric charge, mass, and so on. This doctrine has had immense success, though quantum mechanics has extended the meaning of atomism and many physical particles are not invariant. But these relatively recent changes are not fully understood, and if one provisionally accepts traditional atomism as valid, it is then a natural miracle that in viable organisms the atoms undergo precisely those relative motions, and the global systems such changes, as serve to maintain all the global properties of the life cycle and its various functions. There may be one in a thousand errors, but if so most of these are compensated.

Even when lying ill in bed, because some organ is disordered, I retain the knowledge that for every process that has gone wrong, millions are still superbly orchestrated. How does this come about? Where is the conductor of the orchestra, or do they do without one? The problem may be to see how a system that is fabulously complex in terms of atomic motions can become astonishingly simple when other more appropriate *organic variables* are used. Until this aspect of the problem has been exhaustively explored, it is silly to confuse the issue by resorting to vitalism. It is a waste of time to get involved in a metaphysical battle; the issue lies in the choice of variables, and on this quantum mechanics has so far thrown little light. When we understand quantum mechanics better we may see that it is trying to tell us what variables to use for complex systems, perhaps even for organisms.

All exact scientific theory depends on the selection of the best variables for any class of problems, or to give it a more technical formulation: *optimal parametrization.* Indeed the main aim in theoretical science is to find those variables which make the subject matter the simplest. The problem of organism can be put in nine words: *Which variables best display the internal coordination of organisms?* Then comes the second, no less important question: *What is the relation of these organic variables to those of quantum mechanics?* Two thousand years of speculation and research are condensed in these two questions, and it may be a task for decades to answer them. Yet if this formulation is correct we are getting somewhere, and the truth may lie around the corner, not hopelessly far ahead.

For we can go a step further by getting rid of irrelevancies and putting the emphasis where it belongs. To do this and to give further precision to this fascinating superproblem of the organism and its organic variables, I have defined what I call the "coordinative conditions"[3] (C.C.) of cellular organisms. This may sound obscure because it is a new term, but it is quite simple and merely defines a task for research. The C.C. are the unknown geometrical conditions which determine the changing spatial relations of all parts, at every level from atoms to organs, in every viable organism—expressed simply, of course, in the as yet unidentified organic variables. The C.C. hold the "secret of life." Life is the development and maintenance of the C.C.; pathology and death announce that the C.C. are no longer met.

When we are at such a fundamental issue it is natural to find that others—as it seems to me—are saying much the same thing in other words. In the science of classical mechanics conditions which restrict the freedom of the parts to undergo arbitrary movements are called *constraints.*[4] Today we have left classical mechanics behind, but the fact remains that in an organism constraints must exist which ensure the maintenance of the C.C., that is, of "life." This has been well expressed by Pattee, here one of the most advanced explorers: "If there is to be any theory of general biology it must explain the origin and operation (including reliability and persistence) of the hierarchical conditions which harness matter to perform coherent functions."[5]

Another aspect of the situation is present in embryology, where Wolpert sees the same problem, and asks, "How does every part know where it is and what it has to do at a given moment, so that the differentiated adult is properly developed?"[6] Wolpert calls this "positional information," and finds that groups of, say, ten cells may for this purpose act together. Pattee, Wolpert, and many others are now converging on one comprehensive issue: what happens where and when and under what rules in an organism and the C.C. assume this to be a *geometrical* problem. The convergence is clear and exciting, and I believe that nonspecialists will be able to follow me intuitively through the next paragraphs.

This variety of approaches must not distract us from the basic issue, which is the identification of the appropriate organic variables that provide the optimal approach toward organic properties. We can now forget the question: *What is life?* and substitute the more fertile one: *Which are the organic variables?* No computer will give us the answer; trained judgment *plus* serendipity are necessary. At this point, as I think science should be fun, I shall impress nonspecialists and perhaps tease specialists by putting on record my own guess regarding some of the properties of these unidentified organic variables. Fifty years of cantering around have led me to this conclusion: the organic variables must be *global* (associated with the totality of the system), *hierarchical* (associated with levels), *chiral* (biased toward right or left handedness, skew or screw), *diminants* (quantities which diminish as a one way process moves toward a terminus). (If this sets any young biophysicist on fire, good luck to him; that possibility is one of my reasons for putting this in print.) In other words, I suggest that global, hierarchical, chiral diminants are necessary to define the C.C. and to display the simple ordering concealed in the myriads of atoms even in a single protein molecule. The C.C. as originally defined are too general; it is only when specific properties are introduced that the conception occurs.

The point is that we have regarded organisms as highly complex, because we somewhat perversely used those variables which could not display their simple ordering. It is not surprising that three centuries of successful analytical methods and of atomic physics have left us accustomed to use variables that do not suit organisms. If I had more space and the reader more patience, I might try to explain that part of the obscurity of quantum mechanics lies in the fact that it is a historical halfway house between the radical and classical atomicity of the past and the more powerful global variables of the future. The history of ideas can be a fascinating drama when one takes a long enough view of one's own century.

Very new problems are sometimes intimidating; one may lack a starting point and a direction. But the approach to a theory of organism in the 1970s has one signal set favorable. It is, undoubtedly timely. The 1960s recognized the importance of *hierarchies;* the 1970s, I suggest; will emphasize *morphic processes,* and, being an incorrigible optimist, I have hopes that an authentic advance toward a comprehensive theory of organism, comprising of course the DNA discovery of 1953 and much else in addition, will follow relatively soon. My optimism has this philosophical justification. *The necessary solution must be as simple and elegant as organisms themselves are,* when viewed properly. What is needed is not further abstraction, still more difficult mathematics, or more years of frustration, but *openness to radically new and promising simple ideas.* What is needed is a subtle readjustment, a new approach to the facts about organisms: in three dimensional space, in the course of one way processes, arranged in levels, biased toward left or right, and undergoing pulsations (see below). The history of science sparkles with totally unexpected miracles of simplicity; this brilliance will not stop today, just because we happen to be alive now. Mathematics has not yet exhausted the elegance of the universe, certainly not that of the organic realm.

My suggestions above regarding the organic variables are not blind guesses. The world view suggests that we look for a hierarchy of formative processes, and we have, now reached a new tentative designation of organism: *every viable living system is a hierarchy of morphic processes in a suitable* environment. (To make this concrete, let me use an idea first suggested by a German mathematician, J. H. Lambert, in the eighteenth century, and put to you this question: Is the hierarchy of the morphic processes in your body in good order? Or are there lesions on some level, and if soon which? Are you sick at some low level affecting only, say, a million cells, or at the highest level in the all coordinating glands and brain? I hope not the latter, which might imply a psychosis, or at least a neurosis.)

But this description of the organism is incomplete; it lacks a necessary feature. Morphic processes, in the inorganic realm at least, normally lead toward a terminus, say a molecule or a crystal, and *stop there,* whereas organisms go on and on, through endless cycles at each level, the life cycle itself usually culminating not in a terminus but in reproduction. These cycles stop only in death. This implies that in organisms the terminal structure does not stay put; something disturbs it, and this means that the hierarchy of morphic processes must somehow generate a contrary tendency, forcing the structure back to its original higher energy state, so that structural cycles are produced, down and up, over and over again, but *down and up along different paths!*

For these cycles generate a continual progressive transformation of energy and materials. The down and up motions pump the energy and materials along one way, rather as a pair of bellows where the valve forces the air one way. But the bellows are not a perfect analogy, because they oscillate to and fro along one path, whereas the organic structure goes around a structural cycle, rather like what in the physics of heat is called a "thermos dynamic cycle;" as in any engine converting chemical energy into work. But here the cycle is in a single molecule or supermolecular structure, not in a gas or liquid.

Many troubles in science arise from having to use old terms when new ones are needed. Several molecular biologists have written of the *oscillations* of a protein molecule, say hemoglobin, between a higher and a lower energy state, as it absorbs and releases oxygen. But this is misleading. The protein molecule does not oscillate backwards and forwards along the same path like a vibrating spring. Let us say instead that it pulsates, moving down along one path and up along another, as the structures of the heart do in its pulsations, between two states, one electrically polarized and containing more energy and the other relatively depolarized and containing less. In an attempt to prevent this mistake some years ago (1955) 1 selected and defined the term *pulsation.*[7]

Every functional structure in an organism when it is not at rest or merely being moved from one spot to another, that is when it is actually functioning, is undergoing myriads of pulsations, perhaps more than a million every second. This is the local or structural aspect of life. Pulsations underlie every function, and when a heart stops pulsating in the emergency ward of a hospital, a bell rings, out comes the emergency box, and the poor heart which thought it was at last resting at its terminus is given electrical shocks in an attempt to start it again around its cycles. This may succeed or it may not.

But here we reach the crucial point. What serves as the emergency box with its electrical shocks for every organic functional structure? What prevents the morphic process, the relaxation of every structure, from reaching its terminus and staying put? What forces the structure to receive energy, to go up around one side of the cycle, so that it can again slide down the other side? *It is the environment of the structure which gives it no peace,* and shocks it into returning to the state of higher energy so that it can then slide back, and so on. and so on as long as life continues. Every local functional structure in an organism is part of a larger local system, and—here comes the important feature—the processes in the larger system disturb, deform, and stimulate the smaller relaxed structure, which thus can never rest but must go on through endless cycles, *because its local environment makes it do so.* This disturbance by the larger system is sometimes, but not always, entropic, in the sense that a flow of energy or materials from the larger system stimulates the smaller one.

We see that the role of the environment can perhaps now be understood more clearly than before. The organism's environment, as has long been clear, provides the necessary energy and materials. But the environment of an internal functional structure does not merely supply what is necessary in the vague non-structural sense of delivering the goods. For it normally delivers its stimulus in a precisely orientated and structural manner: it forces the relaxed structure to follow a particular path up to its higher energy state, so that the structure can once more relax along its own different path. This is not fully represented by (scalar) undirected energy or entropy, but involves precisely orientated (vector) changes of structure in space. We are in a pervasively patterned world where the significance and effect of every quantity lies in its direction. We are forced to accept the fact that the process which occurs in every larger system can generate a disordering effect *on* its part system. There is a quasi-dialectical interplay between each structure and its immediate environment; it needs its antagonist if it is to keep repeating itself.

To express the idea anthropomorphically: the ordering and disordering tendencies cooperate to sustain life. *Life is the co-operative interplay of ordering and disordering processes on pairs of levels in the organic hierarchy.* This particular formulation may be faulty or not, premature or timely, new or old that does not matter, for *the idea it seeks to express is valid. My* decades of meditation authorize me to say that.

The fascinating thing is that, given a favorable condition, life never gets tired. Of course, pathology and death are always at the door. But pulsations, when allowed to, go on and on, and the products of the ordering or morphic tendency are accumulated, while there are no persisting or accumulating consequences of the disordering tendency. The entropic tendency merely disturbs; while the morphic builds up structures, replicates them, develops them from egg to adult, and reproduces them: So the morphic tendency has a remarkable privilege over the entropic: it leaves cumulative records, while the entropic has nothing to show for itself, being merely a randomizer or disturbing agent. One might almost say, misusing language slightly: the morphic tendency leaves a memory; the entropic doesn't.

The role of the world view in this provisional approach to the organic realm is not to provide definitive solutions of particular problems, but to clarify the formulation of problems so that solutions may more easily be found. It also serves to throw light on present ignorances. Here are some examples:

1. *Can a living system with all its properties be described in terms of physical laws?* Countless scientists have taken up dogmatic positions on this question, either as reductionists or as vitalists. But both positions are premature, for we have not yet got a unified theory of physical forces, nor has the geometrical character of organic coordination been identified. The crucial question is:

   Are the unified laws of physics logically rich enough to describe coordinated organisms? This cannot be answered until we know the unified laws of physics *and* the coordinative conditions of organisms.

2. *Is the organism a machine?* If a "machine" is defined as a spatial arrangement of invariant parts obeying already established atomic laws, the question is misleading. The organism probably requires global not atomic variables, but as these have not yet been identified, we cannot assert anything regarding the relation of these organic variables to the traditional atomic ones.

3. Another topical issue is: What is the scope of molecular biology? Or better: *How far can the quantum mechanics of, say, 1930, and the chemistry based on it, go toward describing the pulsating structural hierarchy of organisms?* A provisional answer against the background

of the world view would be: Not very far; until an extended quantum mechanics and a strengthened biomolecular theory identify the global variables characterizing the organic hierarchies.

4. This suggests another problem: How *many steps of increasing complexity of structure can be identified in biopoesis* (the emergence of life) *under the morphic tendency?,* How many structural levels are there in the simplest cellular organism? A theory of biopoesis will one day give the answer. The precise number of steps is not known, but it must probably cover (a) the basic chemical constituents of all organisms, (b) polymers replicating by pulsations, (c) hierarchically differentiated systems capable of growth and reproduction, and (d) the stabilization of units (cells) capable of replication by division. At a guess, the complete theory of the emergence of cellular organisms will probably require some five or six separable steps.

It is agreed that at the present time some of the more important ignorances in biology are morphogenesis, the working of the central nervous system, and memory. But these are best interpreted as aspects of a single comprehensive problem, namely, to discover the character of the coordinative conditions which determine all organic properties, of which the following are primary: the differentiated growth of functioning organisms from the fertilized ovum to adult, including all differentiated functions, fertilization, pathology, self-repair, and death. We have seen that one crucial question is: How does a unit at any level know where it is and what it has to do at any particular moment? The coordinative conditions must answer this in the most general and clear manner.

One of the subtlest and most fascinating features of the conception of organisms which the world view suggests is the presence of what may be called the *vital surplus.* In general the morphic tendency generates stable structures, which may or may not be new. From haphazard unstable initial conditions it can generate a relatively stable terminal form which may or may not have existed before. This is of great importance in the inorganic realm and in biopoesis. But in the organic realm it acquires a special significance, where it produces what will be called the "vital surplus." This is not the same as Bergson's "élan vital," as it is of universal origin and does not imply any discontinuity between inorganic and organic. The term is appropriate because the vital surplus results in effects which are *more than adaptive.* (Here I follow ideas expressed by Jean Piaget.) The presence of this vital surplus implies that the total processes in organisms are not merely life preserving, but life enhancing, not merely adaptive but formative and sometimes creative: Thus to mention some of its effects:

1. The morphic tendency molds new stable; internally adaptive *favorable mutations* from haphazard unstable changes in the genome. (See chapter 2, note 8).

2. This tendency forms residual (*memory*) structures which preserve a record of the process which formed them.

3. In the human brain it develops from such records new unified forms (*induction*). (See chapter 4, note 4c.)

4. In the human brain it develops higher order unified forms (*creative imagination*).

5. It establishes in the human brain structures providing mental criteria for activities promoting unity (in an individual, in pairs, or in groups). These criteria are known as *values.* To take an example, charity (*caritas,* love) is more than adaptive; it is an expression of the vital surplus in man, passing beyond merely adaptive or utilitarian criteria. The core of the ego cannot be narrow self love; the vital surplus in man flows into his relation with others.

But are these mere speculative generalizations, lacking any constructive novelty, any biting edge which can cut through old problems and open new vistas? What is the crucial logical and scientific novelty in this approach?

I have already stressed that the world view is global and finalistic, and these aspects can be combined into one formulation: the world view—in the appropriate realms—implies the use of "global diminants"—quantities characteristic of a total system which diminish as a structure approaches symmetry, deformations which disappear as the structure relaxes. This is the novel biting edge. If universal laws are found employing global diminants instead of local invariants, many problems of the organic realm are simplified.

It would be a mistake to undervalue such tentative ideas, not yet supported by specific empirical applications. Few doubt the value to intellectual understanding of the neo-Darwinian or synthetic theory of evolution by natural selection. Yet I believe that in no single historical case of a major evolutionary advance have both the exact favorable mutations in terms of molecular structure and the precise environmental factors responsible for selection in that historical period been identified. (See chapter 2, note 5.) For this and other reasons the theory of evolution, though a most useful stimulus to thought and experiment, is not an empirically confirmed "hypothetico deductive" theory in the standard sense of one current school of the philosophy of science. But this does not lessen the restricted value of the theory. Similarly the vital surplus is a powerful aid to clear thinking in biology.

We saw in chapter 3 that the ordering and disordering tendencies are intimately related. In organisms the ordering tendency needs its apparent antagonist in order that cycles may be completed, and the life processes continue. Life, if interpreted as the ordering principle, needs its antagonist if it is to emerge and to persist. In a fundamental analysis life is not a Hegelian dialectic, if, as we have seen, the two tendencies are not basically opposed, but represent the different forms taken by a single tendency in contrasted systems. At all the necessary times and places the ordering principle in particular structures opens the way to its apparent antagonist, which is the same tendency but in a larger structure.

The view of organisms presented here can be clarified by a further consideration of the status of the current neo-Darwinian or synthetic theory of evolution. It appears to many students of evolutionary problems that the current standard formulation of the neo Darwinian theory is inadequate in relation to the following aspects of evolution:

1. *Macromutations*, or alterations in the hereditary structures associated with major favorable changes in the internal organization of adult types.

2. The role of *internal factors* as an additional influence restricting the possible direction of evolutionary changes.

3. An evolutionary step of the highest theoretical interest and of special significance is the emergence of our own species, that is, a species with a capacity for *articulated speech* or true language. It seems that no subhuman species can learn to use vocal units capable of potentially infinite alternative combinations. This may correspond to an important step in the internal organization of the central nervous system, mainly in the brain, dependent on a macromutation generating this important novelty. If this is so we are at present almost entirely ignorant regarding (a) the precise chromosomal changes involved, (b) the part played by the molding effect of internal coordination on these arbitrary changes, and (c) the role played, if any, by standard external factors such as competition, predators, new habitats, climatic changes, and so on.

Here a major difficulty is that the use of vocal communication, for example, by speech, leaves no direct physical record in the form of identifiable objects in geological strata. But it is possible that, when more is understood, it may be established that the evolution of language occurred in parallel with the systematic manufacture of tools, or with certain types of cave drawings; or other persisting records. It is reasonable to expect advances here, because only relatively recent historical periods are involved, and because it is of such outstanding theoretical and human interest.

My guess—and it is no more—is that this step, the emergence of language, is so fundamental that it may well be associated with a basic change in the internal structure of the genetic system responsible for the higher neural organization, that basic change resulting from the internal molding into a more highly unified and stable mutation of relatively arbitrary genetic changes due to random causes such as radiation. Random and unstable transformations cannot alone be responsible for significant changes of internal structure; structures which are stable and conform to the structural environment must be molded from them.

I repeat, this is speculative, but it suggests an interesting heuristic principle: *Internal factors, molding arbitrary genetic changes into new unified stable forms, may be responsible for some or all of the still mysterious favorable macromutations.* Until such mutations are better understood evolutionary theory is incomplete.

6

Man Today

Man is a gregarious, language using,
unity seeking organism, the human
imagination being the supreme ordering
agent in the known universe. But in this century
he has to overcome his dissociation

> In fifty, a hundred years from now the descendants of these two will act and feel worse than we.... Such boys are springing up amongst us .... They seem to see all the terrors before they are old enough to resist them . . . . It is the beginning of the end of the wish to live.[1]

Thus wrote Thomas Hardy, in 1895, echoing Schopenhauer's negation of the will. These words were one of the many signals from the mid century onwards that the brief century of Antiman was about to begin, a deeper plunge into nihilism than the West had ever known before, for it no longer trusted the fictitious assurances of transcendental religion.

The better to understand Antiman and ourselves we will take a brief glance into the past, remembering how treacherous are condensations into a few sentences of a complex and changing social condition. While the German and English romantics had expressed their radiant belief in man, the most influential writers of the nineteenth century displayed, beneath any frail hopes, a basic pessimism repudiating contemporary man as a creature of conflict without a future. Among many others, Schopenhauer, Kierkegaard, Hawthorne, Melville ("For all men who say *yes,* lie"), Dostoevski, Marx, Nietzsche, and later Freud, Spengler, Picasso, Kafka, and Beckett, announced Antiman, the man who could not bear his emptiness and his conflicts. Nihilism, defined by Jacobi in 1799, had become by 1870 a manifest threat to European society. Nietzsche was, a yea sayer, but to life, not to contemporary man. Being artists, many of these expressed a complex attitude, but most of them stressed the anxiety, absurdity, and tragedy of human life, and scarcely concealed their hatred for what being human seemed to mean. Beneath any superficial favorable features which they saw, these representative figures all painted contemporary man as essentially hateful, doomed to conflict and distortion, conceivably to be replaced some day by a new and better type. Wells saw man at the end of his tether, and Gandhi sounded a retreat. Antiman prepared to destroy himself, and technology overreached itself. All the signals were apparently set for doom.

But we should see this in perspective. In spite of this emotional reaction from a more optimistic mood, major long term social advances were being achieved. As Polanyi says: "The past three centuries have seen an outbreak of moral fervor which has improved modern society beyond the boldest thoughts of earlier times."[2] This is true, and it makes the recent disillusionment of Antiman all the more striking. There is here a clear separation of individual judgment from social action. The community was moving ahead, but the individual despaired. In spite of the continuing social advances, perceptive individuals have been in despair, and with some reason, for the forces driving society were increasingly technical, and the disasters of this century are without precedent. Today we are cynically learning to live in the shadow of the bombs amid increasing violence, doubly damned, as it would seem.

Some claim to believe that nothing changes in history. The truth is that there have been local alternations of creative vitality, of pessimism, and of destruction. But has any earlier generation had to suffer during one adult life of threescore years and ten, as my generation has, anything to equal two world wars and many minor ones, Hitler, the camps, Stalin, Hiroshima, Vietnam, and the arsenals of the bombs? Violence *plus technology* is new, and calls for an unprecedented response, if there is still a will to live. I give Antiman a brief century, say for convenience from 1880 to 1980. He cannot survive much longer, as his obvious aim is self destruction and he now has the power to achieve it. His aim is his one admirable feature: he knows, rightly, that nothingness, sheer nonexistence, is preferable to nihilism. Nonlife is better than life repudiating itself. The individual is ready and happy to die, as we see today on every side. In this he is certainly preferable to the indifferent subhumans lacking charity whose only concern is to survive, with whatever comforts they can grab, until violent anarchy takes over. The widespread adaptation to the existence of nuclear arsenals as an accepted fact of life is one of the most alarming symptoms today, though I believe that the less conscious levels of the mind are preparing a new response.

But violence proves vitality, and where there is vitality there is hope. In the heart of Antiman lies the protest: Need human life be like this?

We do not have to study primate behavior to understand the value of violence for man today. Leaving aside violence as the direct expression of competition, I see the following factors as generating exceptional violence today. (1) Biologically, as an animal organism, man needs stimulus, excitation, sensation, excitement; where religion fails, violence gives him this. (2) Humanly, as a person with feelings, Antiman needs the orgasm of violence to inhibit his unbearable despair. (3) As an active member of society Antiman uses violence to express his resentment against life as he knows it, and to achieve results by frightening authority. (4) Fifty years of officially sanctioned violence in major and minor wars have taught him to enjoy the ugly fun of the gun. There is no need to look to hypothetical theories about hereditary tendencies to aggression when the truth about violence today is so clear.

It is as certain as anything ever is in history that violence will increase until either all civilized order collapses or a new generation with a will to live takes over. Consider your world leaders one by one; measure them against this human situation; and ask yourself: Can they possibly lead mankind out of this mess? No; only a new generation can. Meantime the churches do not unite; the West cannot establish a stable currency; only the international technocratic community thinks it knows where it is going; and the privileged are silent. Behind all this is there or is there not in mankind or in the West an authentic will to live? Whether the bombs are used or not, their existence will within decades transform human thought, either by destruction or by forcing man to think anew about himself. He cannot avoid discovering what he feels and thinks in this new situation.

Let me make one point clear. The will to live, the joyous acceptance of life, does not imply blindness to human suffering, others' or one's own. Nor does the experience of moments of joy eliminate long dark nights of the soul. Romantic optimists and knowing pessimists do not express any absolute or objective evaluation of life, which notion has no meaning. These two attitudes and all their variants express merely the momentary condition of an individual, his relation to his circumstances. But while the nihilists can always destroy, the optimists have a powerful biological factor on their side. It is part of man's inborn nature to struggle to overcome whatever limits his actions and his coordination. The vital surplus in him drives him to create, to invent, and to open up new vistas, and brief moments of joy enable him to survive the long hours of darkness. I have said (in *The Next Development in Man*) that thought is born of failure. The vital surplus in man, confronted by his recent failures, will lead him to think how he can overcome these. If there were not this creative spark in man, enabling him at times to find a splendor in human life, its beckoning sense of a latent perfection awaiting him and its flashes of joy, he would never have created the noble, life enhancing elements in human culture. There is a vital surplus in man, and it is on the side of what man calls "good." In the next chapter we shall see one way in which this vital surplus operates through decades and centuries.

Nothing is harder than to identify major historical changes while they are occurring, in advance of their culmination in characteristic new ideas or institutions. Yet it has long been evident to me, and perhaps is now to many, that during this century a major metamorphosis of the Western psyche has been taking place. I have analyzed this in *The Next Development in Man* and shall therefore only summarize it here, though it is of primary importance for the understanding of our time. During the centuries after 500 B.C. a dissociation of the Western psyche developed—which I called the "European dissociation"—the main social function of which was to stabilize family and community life and property against the challenges of the aggressive threat of the male sexual instinct. In this dissociated state a prior more harmoniously integrated condition was replaced by a condition putting the responsibility on one aspect of the psyche for controlling the remainder. This dissociation took many contrasted forms, though they were all in some degree correlated and they cooperated. Thus belief in a separated *God* was expected to control the remainder of man (what we normally call religion is the operation of a partial dissociated substitute for full organic coordination), money to determine personal and social activities, loyalty to the community to do the same, *rational ideas* to control behavior, *scientific inquiry and technological aims* to overrule all else, and so on. The various doctrines supporting these factors had this in common: one component in human nature was separated from the rest and made responsible for controlling individual behavior in the interests not only of family and community stability, but also, in most cases, of a privileged group who benefited by this situation. All the European communities during the last two millennia have been stabilized by dissociations of this kind in which two or more elements cooperated to maintain their joint dominance. The age long cooperation of Christianity and money is obvious, and ugly.

The unique feature of the twentieth century which distinguishes it from any previous period after 500 B.C. is that the age of communities based on a dissociation of the psyche is coming to an end. This can be expressed even more sharply. In the years since 1914 the multiple dissociation of the Western psyche has collapsed. God is dead, the abuse of money has become intolerable, nationalism is endangering the species, rationalism is clearly powerless to control human desires, science and technology must themselves be controlled. The old gods of the dissociated era have been shown up and have lost their magic and their efficacy. They are ignoble, not noble. In their place, it is increasingly understood from immediate experience that only nondissociated, spontaneous, coordinated human feeling, thought, and action—with its accompanying joy—is self justifying. The old ways are now known to be rotten to the core, antiorganic means to undesirable ends, and there may be just time to promote Unitary Man before Antiman commits his justified suicide.

Man longs for certainty, but that he cannot have. We cannot know, and should not try to know, the future of human society. No man has any right to assert: We will come through, or even: A favorable outcome is probable. Probability is irrelevant to history; there exist no measurable factors determining what our fate shall be. We can only know what is possible.

But if we cannot have certainty we can at least have clarity, the redeeming clarity of the world view of man in nature, years or decades before the sciences give it their indirect confirmation. However, to accept this, man must have been purged of his prejudices such as being in an extreme degree pro- or anti-science or religion and be prepared for fresh hard thinking, open to the new everywhere. He must use his imagination, blunted as it has been at the deeper levels by technology and industrialism. Moreover, he must accept this fact: *Clarity carries no guarantee of any kind,* of survival, of success, or of benefits. Clarity is an aesthetic, not a utilitarian value.

What, then, does the world view imply for man's conception of himself, his feelings, his thoughts, and his actions? This above all: morphic man knows that every life enhancing impulse in himself is an expression of the organic tendency toward coordination, itself an expression of the universal morphic tendency. This awareness runs through many levels of his mind. At one level man experiences freedom of choice; he feels himself to be a free agent seeking order and harmony. But at a deeper level, aware of more, he knows himself to be less than free, the instrument of forces greater than himself.[3] There is no contradiction here. The old antithesis of free will and necessity vanishes in the hierarchical view of man. The "higher" levels of the mind express more specialized factors, the "lower" more general. At one level he experiences freedom of choice, but when he becomes aware of the deepest level of all, he loses free will and experiences the bliss of enjoying and serving a pervasive unity. For joy is simply vitality without discord.

In this book I am opening a new vista. By achieving this clarity regarding the morphic at all levels, human reason turns a corner, and man sees what he has never seen before. He sees—when he is not pathological—that there is no agency or power in him, no desire, thought, or action, no healthy process of any kind, which is not the expression, direct or indirect, of the tendency toward organic coordination active within himself. His supreme blessing comes not from God, but from the fact that he is an organism seeking coordination. For example, all the operations of his brain, express the self regulatory tendencies of the animal organisms in their most general form.[4] To put the same fact in other terms: no religion, no philosophy, no science, no medicine, could assist man by one jot were this underlying organic tendency not present and active within him. In the most austere, authentic, and objective sense, recognition of the status and power of the morphic processes at all levels in himself carries human awareness beneath the perceptions of the traditional religions and of the special sciences to an aesthetic level hitherto experienced and expressed only by poets, for several poets have said what I am trying to say here. To experience, this vision is joy.

The romantics believed in man, and they knew that the imagination alone could fashion man anew. In this they were right. But they lacked the historical sense which could tell them when and why man had first failed to be harmoniously himself, and when and how the imagination would be able to restore him to himself, to bring back his sense of harmony. Today we know the historical source of the dissociation and we are experiencing its collapse.

There is much still to be learned about the history of man, for example how the remarkably rapid development of language came about, the relative roles played by unconscious and conscious factors, and much else, but the etiology of human conflict, the source of the split in human nature, for me no longer presents a problem. The explanation I gave in 1944, in *The Next Development in Man,* still stands—I am more sure of it now after thirty years—and I need not analyze it further here. Whether my formulation is faulty or not matters little. The point for the future is this: the development of the human mind has now reached the stage where it can begin to understand not only its own failure, but how to repair the damage: by cultivating immediacy and spontaneity and by seeing the human person and his mind as a potentially coordinated hierarchy of morphic processes. Man cannot be perfected, but deep lesions can be identified and removed. The individual is not to be understood by ever deeper pseudoprobing of his own subjective mind, using already outgrown psychological theories, but by letting in the fresh air of sanity and clarity, dropping the prejudices of old doctrines, religious and psychological. The most influential theories were all misleading, for they neglected the fact that man is a potentially integrated organic being with a sense for perfection.

The astonishing and glorious fact is: *Man benefits by discarding God!* The redemption Christianity dreamed of is gained by discarding its mistakes. Dostoevski, through some of his characters, wrongly suggested that without God and the belief in personal survival after death all things would be permitted, though he did not always maintain this. Many cultures and countless individuals have shown a satisfactory harmony and stability without transcendental beliefs in the Christian sense. A morality sustained by the fear of God, or by belief in rewards and punishment, is rotten at the core: it reveals the dissociation at the root of human nature in that period. We must each find the inner voice for ourselves. Here the official Christian doctrine is a drag on the human spirit, a social scandal which must be stopped, a utilitarian poison measuring virtue by benefit, a sectarian crime. Man must act more from whole naturedness, less from hope or fear, never from a sense of guilt or sin.

The yearning for order and unity in man biased him toward God; he was glad to accept the existence of a principle of superhuman excellence. Now we know that it is not superhuman. Man could never conceive a quality which it is not within his capacity to experience. Aspiration is the sign of a capacity, as Pascal knew. Man is capable of an excellence "out of this world," above that of his normal everyday life. In his desperate plight today man has to put this most powerful part of himself back into its place, as an integrated element of his own organic and personal nature. Nihilism pervades the world because he is only halfway to clarity; he has discarded the pseudo-God but not yet truly honored in himself his own faculty for experiencing joy in perfection.

That part once projected into the heavens as a pseudo-God is no more and no less than the human imagination at its highest, for it seeks, creates, and enjoys unity. Read Shelley's "In Defence of Poetry." It is all there, together with a few mistakes, easily neglected. Or turn to Goethe, in whom, beneath all his pomposity and fear of surrender, I found the best exemplar of Unitary Man.[5] Or if Goethe seems too far from our problems today, take a more recent figure—I will not say a lesser one; the same truth is at the core of all three—Herbert Read,[6] whose faith in the power of imagination never wavered. There are probably many others, religious poets or philosopher idealists, who would serve as well, but I have not found them.

In this transition from God to the human imagination and sense of perfection, one theme requires a fuller development than I can give it here. The original source and core of all religions, without which they would be mere systems of ethical precepts and rituals, was the experience of transcendent joy, sometimes called "mystical," though that term acquires its meaning not from any definition, but from the experience of those relatively few individuals who tried to put it on record. This transcendent experience of joy can come to nonbelievers and can be given a natural interpretation. No task is more urgent. We need to know how widespread this experience is or can be, what role it serves in the life of the individual and the community, and how far it can help us today. I have not found any adequate naturalistic interpretation of mystical experience,[7] though one may exist.

So we come back to the imagination, here regarded as the primary capacity of the human mind of which all other mental capacities should be understood as special cases or applications.[8] It has been called the poetic imagination, the creative element in the mind, the inventive faculty, and so on. But in most studies of the imagination, leaving aside Goethe's *aperçus*, one factor is missing. The imagination depends on the interplay of at least two levels in the human mind, one more conscious, one less. There are moments when a new term can aid, not confuse, our thinking. I shall therefore call *perconscious* all those mental processes in which two or more levels are at work, one conscious and one less so. All imagination, creation, conception, invention, and induction is perconscious. Pure logical inference, mere calculation and computation. is not perconscious; it tends to remain on one level and leads one into the barren realm of quantity. As Shelley said, "Our calculations have outrun our conception." The world view brings to man fuller awareness of his own creative perconscious; he may thus be able to escape his obsession with quantity.

The perconscious pervades human life. It would take a polymath of unique range to trace its role in the creation and maintenance of cultures. I will take only one example which throws light in diverse directions. Many of the psychological and literary critics of this century do not seem to have understood the facts for which the term "perconscious" stands: the omnipresent interplay—except in sharply dissociated natures—of several levels of the mind. Consequently a story or poem is said to be allegoric of one particular aspect of experience, a human action to reveal one motive, and so on. In fact, nearly all human motives, actions, creations, and symbols are perconscious, expressing many levels at once. For example, Herman Melville had read widely, and much of his writing contains obvious allegories. But this does not mean that he deliberately constructed a story or an episode to illustrate only one universal aspect of experience. At his most creative many levels were at work.

Critics spending their lives thus interpreting symbolisms, can never explain the unique imaginative power of the greatest writings, which cover many levels from the deliberate "logical" actions of daily life down to the all pervading "mystical" sense of the universal which underlies everything else. Genius is the perconscious come to full power. Logic and mysticism can achieve a happy union in imaginative individuals as aspects of one perconscious creative personality. But we must leave literature, and return to man.

This, awareness of the perconscious involving several levels of morphic processes implies a radical revolution in social attitudes and opinions, but no reliance on revolution by force. We are already so far advanced in this rapid metamorphosis of the psyche that no political or economic planning can keep pace with it. No reliance on quantitative measures or statistical trends in any human realm can be of value over a decade or more *if man changes his mind,* and he is doing that now, at an unprecedented rate. We are at the threshold of mental instability. At the present time, painful as the fact may be to many well meaning but naïve planners, no important predictions about the future of any community are likely to be reliable more than ten years, because man is now busy changing his mind. (Take any major prediction and try this out.) Those who hold political power are now so baffled by the absence of steady foundations and reliable trends that they do not know where they are going (take any president or premier and watch him switching policies every two or three years in an attempt to keep up).

Man is certainly now changing his mind, and in many communities both the rate of change and misinterpretations of the direction of change are causing alarm. As I see it, the movement which has now been evident for some decades is not fundamentally anti-intellectual, or against reason, or neomedieval. It is arational, seeking to look beneath reason, not irrational, or contrary to reason. It is certainly against those forms of reason which have proved harmful: overabstract reasoning, overreliance on traditional principles, and the worship of the mere logic of techniques. But the revolt is not against reason or even science if these are properly adapted to the new context. The revolt has explored the realms of emotion and of paranormal experience in order to discover what human capacities actually are. It is my conviction that this movement is leading toward a strengthened conception of reason as an indispensable instrument of the coordinated human organism.

In one realm I believe that some degree of cautious prediction is possible: that of basic values. The supreme human aim was once supposed to be "the greatest happiness of the greatest number." Perhaps this was never meant fully seriously; certainly happiness cannot be measured, nor can the happiness of individuals be pooled. This conception is plain silly, and only an age obsessed with quantity could have invented it, though it is still being used by philosophers. Some prefer "fulfillment of the individual" or "optimal self-realization," but these are also empty terms which beg the question.

Since it is better to be clear and specific even if others may not agree, I will state my personal view. I suggest that under the world view one value becomes supreme: *joy*, for all; an unsought state of grace in which life is enhanced and the imagination surpassed; the unexpected evocation of one's fullest aesthetic response; the bliss which transcends the personal and the purposeful; the surprising experience of perfection; the awareness of being alive without discord. Some experience joy in great intensity and purity; others, in daily life as it is for most, may know only moments when what is mere pleasure seems for once to surpass itself and to take one by surprise. This capacity for joy is given to man in his heredity, and only a dissociated culture can take it away. Persistent suffering cannot be borne without moments of joy. Overstressing one aspect, Wilhelm Reich called this "the function of the orgasm." His point is right; his formulation too narrow. At the root lies this principle: man should not die without having lived, without having experienced enough joy to bring him peace when the end is near.

It may be harder today, perhaps harder than ever before for many, to see and to enjoy what there is good in human life. In earlier times human tragedy was, both as a matter of deliberate policy and unconsciously, hidden away from the sight of many, or of the more fortunate. Today the facts of war, violence, poverty, and disease, of nationalism, and of racial and religious intolerance are advertised and brought to our attention daily. In view of this some may tend to take for granted, and to forget as great human achievements, the rising standard of living (for some), the benefits of modern medicine (for some), the release of sexuality, and all the achievements of science, which for many remain, intrinsically valuable in themselves. But science is now socially tarnished, and the urgent problems of the human community override all others. It is certainly a time when to perceive the good requires an active quality, a vital surplus, in the person.

The values which follow from joy are not moral, but aesthetic: openness to experience; respect for the uniqueness of the individual; tolerance toward variety; unification of emotion, idea, and action; immediacy and spontaneity in experience and action; the promotion of a universal human consensus recognizing these values; and perhaps most fundamental of all and primary to all these others, what I have already implied: *the cultivation of the joys of the imagination,* in the deepest and fullest sense. Need I add that all this involves a critical supervision of arbitrary technocratic power and a further reduction of arbitrary money power? But all particular social and political aims are hollow unless the current nihilism can be replaced by a positive ethos and a new will to live, a new joy in a new generation.

The imagination, as conceived here, expresses the vital surplus, passing beyond adaptation to the past and present, to the construction of the future. But this very surplus heightens our awareness, our tension, and our fears. If we are vital, how can we escape this doom, this vicious circle of awareness and tension? The very intensity of our will to overcome our human failure heightens our struggle and paralyzes us. Here the perconscious offers the only path of escape. *We can will to relax our will.* It is well known, particularly in the East, that this is possible. The powerful will of the West is destroying us as community and individuals. But it is open to each of us, at a deeper level, to say to the Ego, on its "higher" level, "You are to relax," and it can, and may, obey. It is much to expect, and the mad will of the West must be overcome not only by mockery, but by revealing to every individual drowned in the current social drift his own utter insanity. He is doing what he does not, at bottom, wish to do. Beneath everything else he longs for a new way to enjoy life He wants to live freely, to discover once again the joy of wholenatured living.

The conception of the perconscious makes it easier for us to see deeper into human nature. In the past it has sometimes seemed puzzling that tragic distress can be a source of aesthetic pleasure. How is it that so central a faculty in man as his sense of beauty can lead him to find satisfaction in human tragedy? The answer is simple. What is tragic at a personal level is beautiful at a deeper and more, general level. A deep necessity, once fully perceived and understood, can bring to personal tragedy a redeeming quality at another level. It is this aesthetic relief which takes the sting from death for a person who unreservedly accepts its inevitability, and the more joy he has experienced in life, the easier this resignation, for he has known fulfillment. This is for me an illustration of what Nietzsche meant by the expression he so often used: *amor fati,* the love of fate. The perconscious mind, through the interplay of many levels, can remain serene amid disaster. We badly need that serenity today.

There has run through these pages like an easy evasion the condition: *provided man is not pathological.* But today he is pathological. So what relevance has my theme? Man is pathological, I suggest, whenever he is not occupied in enhancing life, his own or others.

This is no empty tautology, for it asserts the possibility of two conditions, health and sickness, which can be distinguished. Sick Antiman I have already considered. There is today the opportunity for Unitary Man to promote health, and more than mere health: the vital creative surplus. Here one crucial fact can encourage us. As a thinking animal organism, man is by his very nature inexorably committed on the side of order, once the path is made evident to him. *The virulence of his pathology will decline, once the morphic conviction spreads.*

Do these thoughts seem to you too far from Vietnam and Bangladesh or any other focus of war and violence? Then perhaps you do not understand that religious convictions, powerful as they have been in creating sectarian unity in the past, have been nonetheless mere patchwork concealing conflict, beside the real power of an authentic world view.

"Very well," you may say. "This may be right. So what? Where do we go from here?"

This question is one I do not need to put or to answer. I am not here drafting a scenario for man. My aim is the expression of a conviction which I believe has power and timeliness. I repeat: These words are not written for a purpose, in order that tomorrow a new social movement may be born visible to all, or a new social advance be achieved the day after. I am concerned only with the present movement of the human psyche, and I write for the countless thousands of others who feel and think as I do and will be strengthened by my words, as I am by my knowledge that those others exist. This is no rosy dream but based on recent experience. I know, as certainly as I have ever known anything, that a section of the young in spirit, humanistic United States[9] has the vitality and vision to see much of what I see, and perhaps even to see it more clearly. Once again, I do not seek to communicate, but only to evoke, to raise by one level in the perconscious, convictions that are already there in you, my responsive readers.

That a supreme synthesis of all positive values is possible now implies no historical magic. It is possible because, and only because, all the authentic values experienced throughout, human history were in fact creations of man's unity seeking organic nature and must, therefore, be varied expressions of *one* value: the morphic nature, of man, and his capacity for joy. To use language that is no grander than the fact: human destiny is continually to realize in new situations, and at new levels, man's creative powers, and now at last he is fully aware that these powers spring from his organic nature, and that there is nothing timely which they cannot achieve. In this clear, restricted, and objective sense *organic nature is on our side.*

Why do I believe that my theme is timely, thirty years more timely than it was when I wrote *The Next Development in Man*? I believe this because I read of new forms of violence every day. This is the terrible irony of historical change, bringing such untold suffering that one cannot bear to force one's attention to it: the old ways must bring man over the edge of disaster—as it seems—before, with his immense inertia, he will accept a radical change in his thinking. This is not a pretty fact, but it is true: the destroyers around us today are the indispensable creators of a new opportunity, beloved antagonists preparing the way.

I have criticized the privileged and the establishments for their blind indifference and apparent inability to understand the human situation. So what do I suggest that they should do? To answer this question is everybody's task, not mine; there are ten or twenty threatening issues. For example, take Bangladesh, perhaps the poorest, most overpopulated, and most miserable corner of the earth, suffering from the aftermath of war, flood, and sectarian hatred, ideal ground for rebellion by force under ruthless, power mad leaders. Is it beyond the powers of our species to transform Bangladesh from a focus of evil to one of good? Certainly it is, if nihilism still dominates.[10]

Are we to drift further into an authoritarian, ruthless, super-technological, power-mad world noncommunity? Here those working in the communications media of the world have a supreme opportunity and responsibility. One hundred men of good will seeking to discover and report the truth, and to spread sane policies, can have an importance a million times their numbers. But only if nihilism is replaced by hope.

I can imagine even the most sympathetic reader thinking at this point: But do you really imagine that some initiative or other new factor can arrest the terrible drift of this machine age toward greater disasters? Why should the next few decades be different from the last? Is there not a continuity, an inertia, in history which extrapolated ten or twenty years forward must bring all human values to nil? Are you not relying on some favorable intervention of an unprecedented kind? Where is our savior?

Yes! I am relying on a historical factor not as yet appreciated and never yet effective in such a degree as it must be now. In full earnestness, I assert my belief that a relatively sudden change is possible, a general reorientation, say in two or three decades. This paradoxical expectation is not silly, because the metamorphosis of the psyche from Antiman to Unitary Man (since I must use these simplified labels) is not a moral advance requiring unprecedented qualities and stupendous deliberate efforts by millions of men and women suddenly becoming morally better. It is an organic and aesthetic metamorphosis, proceeding of necessity once the conditions are given, a transformation already long prepared and guided by a powerful factor, as we shall now see.

7

The Unconscious Tradition

Our conscious aspirations are aided
by an unconscious tradition

An idea may appear to burst suddenly into the awareness of an individual or a community, but the full implications of a powerful new conception are only slowly understood. Some of its more important applications may not be widely appreciated for decades or even centuries after the idea was first formulated in bare outline. One might think that the conception of the unconscious mind was by now fully developed. Some hundred philosophers, poets, psychologists, and doctors had considered the idea of unconscious mental processes since 1700[1]; Freud's great work, mainly from 1895 to 1930, called wide attention to it in the Protestant West, and his contribution and Jung's have now been under discussion for several decades.

Yet it has gradually become clear to me that one of the most significant applications of the idea, of great importance for mankind today, has scarcely been noticed, though one thinker came close to it a hundred years ago. This is *the role of the unconscious in history* and the fact that behind the conscious tradition, of which we are aware, there is an unconscious tradition, which plays a highly significant role. To put it in a nutshell, while the transcendental God is bogus, there is a genuine unconscious tradition continually at work in *our favor,* unlike God, who, if he exists, prefers not to do anything. Moreover, as we shall see, while the conscious tradition varies from one community to another and tends often to divide the race, the unconscious tradition is biased toward universality and unity.

I shall take for granted the ordinary conception of unconscious mental processes. There is no doubt of the continual occurrence in our minds of processes of which we are not aware while they occur.

When the thought suddenly comes to me at a time when I am otherwise occupied, "Oh! I've forgotten to do that," the postulation of a temporal continuity underlying such sudden thoughts suggests that there has been a misfit between two mental patterns at an unconscious level. Perhaps it is between the engram or lasting trace of an intention and the engram of actions already taken. This misfit generates, or itself constitutes, a mental tension which, when conditions are appropriate—this usually means when we are relaxed, not occupied in deliberate activity requiring our concentrated attention—suddenly draws the attention of the conscious mind to it. For such an influence of one level on another always involves the passing of a threshold, which is necessarily a sudden event, and the new idea therefore seems to have emerged instantaneously. The terms "tension" and "attention" come from the same root meaning "to stretch." The misfit stretches the mental structures at one level, and this, when conditions are favorable, provokes attention to it. Thus the conception of unconscious mental processes arises partly from the demand for a causal continuity which can help to account for sudden awareness or recall, memory itself being one aspect of the unconscious mind.

It has long been known that problems can be put away out of conscious attention and left to mature at an unconscious or less conscious level. The fruits of, an unconscious process may come to us later, and there is no doubt that the unconscious mind can "work" on many problems simultaneously, like a multichannel, link, at least if they lie in different realms. This needs no argument, the experience being commonplace. As I said, the suddenness of a new awareness means that a threshold has been passed, and Piaget found in the development of the child mind that the clarification of a rational structure is achieved in a sudden step, often accompanied by a simultaneous formulation of aesthetic and ethical criteria.

We now come to a more complex notion, and for convenience and brevity I shall here continue to use the terms "conscious" and "unconscious," though they are misleading. Unconscious mental processes do not constitute one indivisible realm, but form a hierarchy of processes, successive levels being characterized by different functions.[2] For example, it is reasonable to assume a highly general or "deepest" and usually unconscious level underlying all mental activities, and a series of more specific levels, more frequently becoming conscious. If this is so, the dichotomy conscious/unconscious is mistaken and must frequently be misleading, because even the deepest level may on special occasions become conscious—for example, as a sense of pervasive unity, bringing transcendent joy—while on the other hand factors that have been normally conscious may fade and for a time become unconscious. I remind you that I use the term "perconscious" as a name for this hierarchy of mental processes when at least two levels are active, the less conscious or entirely unconscious processes being inferred from conscious features.

An individual of exceptional imaginative power is one in whom many, perhaps all, of the levels are in continuous or frequent free interplay. As Goethe knew from his own experience:

> was von Menschen nicht gewusst,
> oder nicht bedacht,
> durch das Labyrinth der Brust
> wandelt in der Nacht,
> "An den Mond."[3]

and he wrote of the "manifold relations between the conscious and the unconscious" in 1832, twenty four years before Freud was born, whose life task was to study, perhaps somewhat more than to cure, those unfortunates in whom there was a neurotic dissociation of the perconscious into two separated realms—inhibited memories and conscious distress—with the pathological result that spontaneous living and free productivity were impossible. Neurosis is the warping of the perconscious through the inhibition of some level, and psychosis is perhaps its breaking into parts—but these classifications are not adequate. Freud mainly considered human pathology; I am here emphasizing the creativity, in man, for it is the creativity in its purest form which we call "divine." However, I am about to suggest in full earnestness that it is more than divine; it can help man as no God ever did. "Divine" is a poor epithet for what I am about to describe.

For now I reach a matter which is not only of great interest to me, and seemingly new, but of great importance for the fate of mankind. If one postulates a causal continuity, as anyone seeking rational understanding must, it follows that unconscious patterns of thought must contain information, or contain features that can later become conscious information, *before* the conscious mind does. At any moment *the unconscious levels are ahead of the conscious* in the task of unifying emotion, thought, and action!

Unconscious thought patterns are ahead of the conscious mind on the path toward unification of experience. Thus the nonpathological unconscious or perconscious leads the conscious! Think what it would mean if everyone could regularly, say once a day, scan the unconscious levels to collect all the new suggestions awaiting exploitation. Imaginative workers do this. Perhaps in a future culture, there will be a routine procedure of mental hygiene: every morning, after cleaning the body, release the mind. Relax, contemplate, scan the unconscious for its daily harvest, and then allow the conscious intellect to do its necessary task of selection, rejection, arrangement, and so on. For except in rare individuals, no unconscious mind is fully well behaved; there are always some scars, lesions, and distortions prejudicing the harvest. Wonderful as they are, we cannot unreservedly trust the unconscious levels, or at least few can; they normally need constant supervision.

However, my aim here is not to speculate about a future culture, but to point to something that has already been happening for some decades, as welcome to man as the increasing violence of recent years is distressing. We are the inheritors not merely of fully conscious verbalized or artistic tradition, but also of an *unconscious tradition* as yet not explicitly formulated or expressed, which is always ahead of the conscious tradition. In every one of us, as we engage in conversation or as we read a journal or book, there lies, beneath our awareness and sometimes dimly sensed between the spoken or printed words, a pattern of new truths awaiting their time to emerge into conscious clarity. This unconscious tradition has been largely neglected by philosophers, historians, and sociologists, though it is a major agent in all important social changes.

The unconscious tradition is the direct expression at the human but unconscious level of the organic coordinating tendency, and it therefore displays a movement toward unity and coordination. On the whole, in, the long run, *there is more pathology in the conscious tradition and more unity in the unconscious one!* The conscious traditions tend to divide mankind, the unconscious tradition tends toward unity, and can heal divisions. The unconscious tradition does not solve particular problems, as does the conscious mind, but continually tends to coordinate. We often know that we yearn for something more than personal, which we cannot identify. This may be a signal that the unconscious tradition is on its way toward a new unification.

We can trace the workings of the unconscious tradition in greater detail. Every parent and teacher more or less unconsciously communicates to the child countless unnamed attitudes, spontaneous gestures, and unspoken thoughts and preferences, all forming part of an indispensable tacit knowledge or craft of living, without which background language itself would possess little power. When something is good and happy and just as it should be, the mother smiles, and by an unconscious mimicry—which I understand that physiology cannot yet explain—the child smiles in response. Here is a primary and supreme value, an acceptance of joy in what is good, communicated without words. It is, sometimes at least, good to be alive. On this mimicry, first of a smile, and then of a spoken word, the entire human tradition rests.

But, more specifically, the parent and teacher are continually conveying preferences, emphases, hints, which cannot at the time be put in words, but are accepted, absorbed, and often mimicked by the child. Indeed even between adults words are often used hesitantly and ironically, as if to say: There is something here that I cannot express properly; you can recognize my frustration in having to use these clumsy words, but perhaps you can guess what I am trying to get at.

This process of suggestion is the essence of language, which is rarely, if ever, as fully clear, rational, or explicit as some thinkers have supposed (Russell and Wittgenstein experienced the agony arising from an overreliance on logic). Language is always hinting at something not yet fully clear to the conscious mind; it is carried by less conscious levels. All language is a hint, with luck evoking in the individual receiving the spoken or written word a corresponding state of awareness. We are all members of one species, by our common heredity potentially capable of understanding such hints and pointers, if we are not unduly warped by unfortunate heredity or experience.

Consider the tradition of the West, say since 1600 when there was a large educated community with a common background of knowledge and technique. In some cases the unconscious processes in an individual culminated in a gift to the community: a new principle, discovery, or work of art. This is genius fulfilled. But in the lives of the vast majority of individuals their private unconscious processes, their daydreaming without full awareness, did not culminate in any definitive new clarity or explicit gift to the community. The majority die without any such achievement. Nonetheless every nonpathological parent or teacher passed on to the next generation a collection of half or fully unconscious pointers: unspoken indications of the direction in which we should move toward the good life or a better way of thinking. Insofar as these partial advances express the universal, rather than merely personal, factors, they promote the advance toward unification. This is the unconscious genius of a community. This unconscious tradition is today for the first time potentially coherent throughout the world, as all share a common background of knowledge, or are moving toward that.

This local, family or school, tradition of unconscious pointers may be difficult to identify, indeed it must be, unless, say, a chain of mothers and daughters kept continuous private diaries over a century or two recording their secret attempts to express their aspirations and disappointments, and their efforts to see life more clearly. But if we turn to the printed word, the unconscious tradition is unmistakable.

Take the file of any great newspaper or journal from 1870 to 1970. For several decades patriotism was universally treated as an unquestioned virtue. Then faint hints arise that this is not so certain, at least not always. What is unconsciously dreamed of deep in every sane man's heart slowly penetrates the awareness of the community. Here and there flashes are evident: some individual finally denounces nationalism as stupid and futile. Then it becomes official doctrine: national wars are out of date—though the nations still behave otherwise. Behind this ethical transformation lies the great dream, originally deep in most, at a less conscious level, and seldom voiced: the potential unity of mankind in no sentimental sense, but ultimately to be expressed in a universal consensus of value and judgment, worthy of the entire species. The tension of unconscious pressures forces this pattern of thought slowly toward the conscious level. In this century, with historical suddenness—in fact in less than one generation, say front 1955 to 1970—this fantastic dream has become an obvious social necessity. The new thought which marks the sudden transformation of the psyche is absolute: *There is* now no *choice!*

This is new, and this new fact can and will work miracles. The newspaper files would show that this explicit recognition has come suddenly, though thousands of writers had hesitantly pointed toward it for more than a century. Today the fully recognized necessity of an effective unity of mankind results from two overlapping factors: the steady maturing of 'a largely unconscious yearning in the hearts of millions, long officially regarded as unacceptable, and the development of the techniques of war to the point at which this tremendous partly unconscious ideal suddenly blazes into full awareness: *There is now no choice!*

The unconscious tradition has for long been ahead of those premiers who claim to promote the separate interest of their own nation. National self interest no longer exists; it is meaningless in the late twentieth century; it points the way to nowhere. But the great idea which the unconscious tradition is today forming in thousands, perhaps millions, of minds is *that the next step is the formulation of a universal consensus*, expressing what men everywhere, because they are human, must accept, unless they are pathological.

This unconscious tradition is distinct from Jung's "collective unconscious," which represents the relatively unchanging tradition of the archetypes of a culture, or of mankind. The unconscious tradition here described is formed by a myriad chain of individuals, passing on from generation to generation their own tiny unconscious contributions to a general advance. While Jung stressed a collective constancy, I am pointing to a slow cumulative development through chains of individuals.

When we say "the time is ripe" for an idea, or that it is "in the air," we mean that the period of unconscious preparation is nearly complete and that a sudden step into conscious clarity must come soon, for the threshold has nearly been reached. It is the unconscious tradition which shapes the *Zeitgeist.*

There will doubtless be some to whom this idea appears too vague and uncertain to be taken seriously. By what right of logic or experience do I postulate that such unconscious processes possess this role in history? Does this amount only to a gradually developing ethical emphasis as the above example may suggest, a matter long obvious to historians?'

What is "obvious" often requires analysis. The role of a slowly developing unconscious tradition can perhaps be seen most clearly in the development of scientific concepts involving no obvious ethical aspect. The Greeks had the idea of permanent matter. Gradually after 1600 this idea matured by many steps, at first unclear and scarcely conscious, but for exact science the essence of "permanence" was eventually found to lie in the *conservation of a quantity.* I have no doubt that historical research would show that many minds were half-consciously striving toward some clarity of this kind, though they could not know the precise form that it should and would take. In fact a great unconscious tradition was leading toward the conception of quantitative conservation, today seen more generally as *principles of invariance,* and, with the usual historical irony, officially recognized at the very moment when they can be seen to be inadequate, since we now need to start with one way processes or tendencies.

In such a process old concepts must be gradually undermined to allow improved ones to develop and take their place. Between the lines of the writings of the greatest scientists one can often sense a half-admitted limitation, an awareness of something lying still ahead, not yet as clearly seen as it will be later. These faint, unconscious hints accumulate, reinforce each other, and help to guide genius to its definitive achievement.

I believe that it is literally true, in a fully objective sense, and will soon be widely accepted, that *since 1914 the unconscious formative processes in countless thousands of sensitive individuals in the West have been developing the necessary elements of a potentially universal consensus,* a future effectively worldwide agreement on what it means or should mean to be human today. Because this hidden formative process has already been at work for half a century, and only because of this, is it possible that a sufficiently universal agreement will be established for civilization to survive? Were it not for this half century of preparation our fate would be already sealed. If the most effective human actions were those in which a deliberate technique is applied to a consciously selected end without unconscious support, there would be no hope.

The collapse of the "European dissociation" of the human psyche is a necessary concomitant of this spread of a universal attitude. For example, the breakdown of the antiorganic separation of "conscious" and "unconscious" has, since the last decades of the nineteenth century, been heralded by innumerable social signals: prophetic anticipations, intuitive convictions of a new human orientation, apparently irrational tendencies actually preparing the way for a stronger foundation for reason. The coming period will be at first marked by the spread of intuitive attitudes, but as the world view develops it will eventually be seen to provide the rational basis for a unification of knowledge and of the sciences and for a policy for man. In fact, if this interpretation is correct, the period ahead will be one in which the tension between intuition and intellect is transcended in a radical unification of all aspects of the psyche. As Shaw said, it will be seen to be silly to be either *pro* or *contra* intellect or intuition, or science or religion.

The existence of this unconscious tradition accounts for an otherwise puzzling feature of the advance of thought and the development of opinion: the strange suddenness and simultaneity in the appearance of new ideas and attitudes. This is well recognized, and I need not document it. Because the unconscious tradition—for example, in the West recently—has been shaped within a common context of science and technology, it is itself coherent, and similar stages are reached simultaneously by many independent individuals. The more efficient the intercommunication within a community, the more nearly simultaneous will be the sudden emergence of a new idea from the shared unconscious background. We can go further. The suddenness with which the new idea finally emerges to conscious clarity in the community reflects not only the fact that the individuals form part of a coherent tradition, but also that in each individual the final emergence of an idea into full verbal clarity always occurs suddenly. Thus the community enjoys its own lightning insights; it passes a threshold and a new vista opens out.[4]

The rapid social spread of a new insight rising from a less conscious to the conscious level reflects the fact that *only the unconscious or semiconscious mind can hold a complex system of ideas stable*; before it is consciously systematized. Thus the growth of a new unification of what had seemed complex must come from the less conscious levels. Similarly the unification of mankind by the spread of a potentially universal consensus can come about only after long unconscious preparation. But the human conscious ego needs a humility alien to it, if it is to accept this fact. Only when we seem to be already over the precipice will the proud Western ego fully accept help from the less conscious levels.

This analysis implies that in a time of convergence toward a new synthesis, many seemingly independent individuals may unconsciously be led toward formulating what prove to be aspects of a single integrated pattern of ideas. The best example known to me, and the most striking and precise, since it comes from quantitative physical theory, is the phenomenon—quite inexplicable if it were not for the unconscious tradition—of the development of quantum mechanics from 1923 to 1929. During these seven years some fourteen theoretical physicists[5] made their own individual and different contributions, though a few worked in pairs, to what was finally found to be a single highly integrated theory. What led them thus, without conscious intent—for *none of them saw the whole pattern until their pieces had been fitted together around 1928 to 1930—to* cooperate unknowingly and yet with perfect orchestration? The unconscious tradition, since we can no longer ascribe such planning to God. From 1923 onwards unconscious mental processes began to press upon the minds of these fourteen men varied aspects of a single melody, with its appropriate counterpoints and harmonies.

This is another parable for mankind. Unless a similar orchestration of unconscious patterns, all forming part of a single unified, indeed all comprehending, pattern lying ahead, has been at work since 1914, I see no hope for mankind. Only the unconscious in individuals can shape a complex unified pattern. God has never aided man to be truly himself; perhaps the unconscious tradition now can and will. I see many signs of it, in spite of, and because of, the present disastrous slide toward lethal violence.

To Western minds, particularly to those who have pondered little on the history of ideas, on changing opinion, or on the emergence of new ideals, this reliance on unconscious processes may seem bizarre. Indeed it must be so, for the time when this factor can be widely appreciated has, I believe, only just arrived.

But if we go to the East, say to India, we can observe a slow social development in the course of history which was, it seems, to a large degree determined by what in Western terms must be called unconscious factors. Without personal knowledge of India, I have long wondered if the Indian people do not live, more than we of the West, guided by factors the significance of which it is not, in a Western sense, fully aware. It seems that I was right.[6] But here one is frustrated by the knowledge that we do not yet possess an adequate terminology. Particularly for a culture such as that of India, the attempted separation of conscious from unconscious factors is not merely irrelevant, but radically misleading. India has lived by the perconscious. Perhaps in some degree China has also until recently.

This leads one to a prophetic hope. It may be that when Europe understands better its own unconscious tradition it will simultaneously find itself in greater sympathy with Eastern life and values. Indeed this is already evident in the increasing numbers in the United States who have begun to study ancient Chinese and Indian thought during recent years. This is one important step. Of contemporary China and Russia I can say nothing, and need not. We have a heavy enough task getting our own aspect of the world view into order. But anything authentic and universally human must ultimately have a universal appeal. This is not a mere tautology, empty of content. It asserts the possibility of an all embracing world view, and that is what matters. But to achieve universal appeal the view must express principles more profound and more widely valid than Marxism as interpreted today in Russia or China. It must have the revolutionary power of a new, deeper, and more powerful conception of man and society.

Reading in sociology I have noted the lack of an adequately fundamental theory of social change. Behind the effects of advances in science and technology lies the influence and cumulative effect of the human imagination, the faculty which shapes new unities, and here the unconscious tradition does the accumulating. It has been suggested that the ultimate urge in the human mind, and the basic fact of consciousness, is the dialectical opposition between subject and object. This is wrong. Constructive thought existed before subject and object were separated, or that separation would never have been established. Hegel is one step only neither the first nor the last word. A deeper and earlier characteristic of thought was simply the ordering of experience That is simple, but in one sense only: it can be formulated in a few words. Its analysis is very complex, for it involves the fourfold interplay of conscious/unconscious and individual/community. A reliable theory of social change certainly requires what is here called the unconscious tradition, for this is the most powerful factor at work. It is the accumulator of imaginative novelties at their origin before they are delivered to the critical light of reason. But we must remember that the unconscious is part of a wider perconscious.

As we have seen, the imagination or faculty for forming new unities is the expression at the human mental level of the coordinating tendencies of the organism, itself one case of the more general morphic tendency. The significant fact, superb for all who care, is that this unconscious tradition of necessity works in the same direction as the human aspiration toward the unification of experience, for it expresses the unifying principle of which these conscious aspirations are the most advanced visible expression.

So splendid an idea is itself necessarily the fruit of an unconscious tradition. In this case the idea emerged already a century back, and pressed its case so well on one individual that he had to write a whole chapter on it. Eduard von Hartmann, in his *Philosophy of the*

Unconscious,[7] had much to say about the unconscious in history, but he did not recognize an essential aspect of the present idea: countless parallel chains of individuals unconsciously passing on, as adult to child or as author to reader through the printed word, an accumulating reservoir of unconscious but potential knowledge. Nor did he sufficiently stress the basic characteristic of all nonpathological human endeavor—the unification of all aspects of experience: thought, feeling, and action.

The main imperfections of my presentation here of the idea of the unconscious tradition arise from the fact that I am compelled—for lack of an adequate theory of the various levels of mental functions—to use the misleading terms "conscious" and "unconscious." Moreover, I have stressed the intellectual aspect, and may seem to have neglected the aesthetic. The truth is that a perconscious tradition is always at work at many levels, and great art is an expression of the perconscious tradition in which the conscious and many unconscious levels are inseparable.

So this is my conclusion. If we come through it will be due to the preparation since around 1914 by the unconscious tradition of mankind as a whole of a universal consensus[8] only now nearly ready to break through into awareness. If this happens it will be relatively sudden, during the next few decades. Moreover, to possess universal appeal it must promote a standard of social justice and of personal liberty transcending the current examples of the United States, the Soviet Union, and China.

The ancient Chinese considered the "Dilemma of the Reformation and the Sage." A disordered world can be reformed only by a Sage. But so long as the world is disordered no Sage can appear. The resolution of this Dilemma lies in the fact, often proved in history, that genius can arise in spite of disorder, pointing a way out. But such a leader would have to carry a psychologically intolerable burden if he were not supported and guided by the unconscious tradition. He could not carry it alone.

8

Conclusion

This book is not only a statement of conviction, but also a historical experiment, since some of the convictions concern the immediate future. The next three decades should show where my thesis is right and where wrong.

The convictions are of two kinds. Those that I consider to be universal in origin I cannot believe will prove wrong. The others are personal judgments, and it would surprise me a little if they were to prove right. For I am an optimist by temperament, in the sense that I look for the good which is conceivably possible, and the great change for the good, which I hope for before A.D. 2000, is much to expect. It is possible, but no more. All personally colored historical predictions are treacherous, and there is certainly no philosophy or science of history which can help us to foretell the future.

The themes of universal origin are the world view and its implications for philosophy, physics, biology, and psychology, which will be great. Here the issues are technical, and perhaps of less interest to the general reader, though if he is under forty, he may well live to see my main theme proved right or wrong if ultimate disaster is avoided.

The main personal judgment is one of great importance. It is that, if total disaster can be held off, there is a possibility of a worldwide consensus emerging soon, and spreading rapidly, concerning what it should mean to be a human being in the late twentieth century, and on steps to implement this consensus. Here we are touching a theme of such social importance that caution is in place. I can only say what I believe. I see the only hope for us in what I consider to be a fact: that an unconscious tradition has since 1914 been preparing the ground for a sudden emergence into general awareness in many countries that such a consensus is both necessary and possible. Moreover I believe that signs of this are already visible.

This consensus will be of heart, mind, and will, and in the intellectual realm it will identify the human imagination as the primary creative capacity in man, the source of religion, art, philosophy, and science. But these terms refer only to particular, mistakenly separated aspects of culture, and they become ambiguous and misleading in a period when mankind is approaching an organic synthesis of emotion, thought, and action. Thus the formulation and implementing of the expected consensus will display some of the characteristics of a world revolution, of a new religion, and of a scientific synthesis. But it will be more than the aggregate of these three. The consensus expressed in action will imply a biological maturing of man through a radical metamorphosis of the psyche toward the global integration of a previously immature and dissociated type of man. Here nothing absolute, or perfect, or unnatural, is implied. What is necessary is a sufficient integration in time. This is much to expect, but not outside the realm of the possible, if one takes a long term view of the biology and history of man and knows that basic transformations always occur rapidly.

One should not underestimate the unprecedented encouragement to the human psyche that the establishment of the world view and the consequent wave of therapeutic scientific discoveries would bring. There will always remain disaster and tragedy for individuals, and no conceivable historical change can eliminate all the causes of human suffering.

But the context will be altered. The structure of the universe at all levels will be known to be less arbitrary than had previously been imagined. A major pattern will be seen to exist in the universe, a hierarchy of morphic processes, a natural plan needing no planner, for at the root it expresses a logical necessity. The human mind will understand its status in the biological realm and in the cosmos, and it will recognize the human imagination as the supreme formative agent in the known universe. One cannot use the language of the past to portray the quality of an experience mankind has never had before; one can only assert its novelty and its favorable character. Will such a metamorphosis of the psyche enable men in many lands to learn quickly enough to live in peace with one another? No one can tell.

I consider that it would be important if, during a decade or so, those interests which benefit from the promotion of violence could be identified, publicly exposed, and brought to trial by a World Court empowered for this task. The elderly and weary will smile at such a suggestion, but the young in spirit might achieve it. In them alone lies the world's hope, and history has not exhausted its supply of surprises. The next step is for man to surprise himself.

Of supreme importance in my theme is the possibility of a swift consensus, which by its authentic expression of what man feels he should be and do in this century makes a powerful appeal to sane men and women everywhere and will be rejected only by the pathological. Now it is a remarkable fact that if the conception of an unconscious tradition is accepted, certain significant consequences follow. The *suddenness, simultaneity* in different regions, consensus of judgment, initial intuitive character, and the *unconscious and unwilled cooperation* of many are then no longer causes for surprise but are seen as historically inevitable features when such a fundamental metamorphosis of the psyche is occurring.

Such a consensus, if it comes, will find many voices stressing different aspects. Yet as I saw it thirty years ago and still see it now, common to all these must be the recognition that what I have called the "European dissociation" is collapsing and that we are entering the age of Unitary Man. G. le Bon suggested that every culture is characterized by relatively few "directive ideas." The period ahead; in my view, will be marked by its emphasis on *hierarchies of morphic processes* in the inorganic and the organic realms, on *man as a coordinated organism*, when not *sick perpetually creating new unities*, new unions of contrasts, and, above all, endowed with the capacity for *joy*.

But this presentation of a possible, future does not cover those features which would surprise me most if they occurred. I have suggested that every thinker should be open to surprises, particularly where his personal convictions are strongest. I shall therefore try to live up to this precept and close this book by listing conceivable events which I do not expect to occur during the next thirty years. Each reader can judge for himself whether he agrees with me or not.

1. I do not expect the discovery of acceptable evidence of the existence, in any sense, of a transcendental personality, God.

2. Nor evidence of the survival of the individual personality after death. ("Personality," as understood in the West, puts the emphasis in the wrong place.)

3. Since the intellect treats classes of phenomena, not unique cases, I do not expect that reason or science will *ever* identify those fac-

tors, if any, which determine what uniquely distinguishes particular individuals. I have in mind, to illustrate a general principle by selecting one aspect, the factors, for example, which make a particular person into an artistic, philosophical, or scientific creative figure with outstandingly original and in some degree timely capacities, perhaps with a sense o£ a personal destiny. In this age, when "science" is supposed by some to be all-comprehending, it gives me pleasure to think that human uniqueness eludes its grasp. But this personal attitude may prejudice my judgment.

4. I now come to a matter where the universal provides me with inadequate guidance and I distrust my personal judgment, because it is a historical matter of such importance that my mind falters when I think about it. This must be a personal frailty, for if I fully trusted my way of thinking, I should have no doubts, since I believe that the theme in question lies within the scope of the world view. So I put this on record for a younger generation to carry forward.

I consider that the possibility of a unification of a physiological theory of brain processes and a psychological theory of emotional and cognitive processes is implied by the world view. Brain and psychological processes both possess a hierarchical character, and a unified theory should be possible in terms of this characteristic common to both. This should finally disperse the most obstructive feature in the thought of recent centuries: the intellectual separation of body and mind. It is difficult to set limits to the benefits that might come to man through such a unification. A rationally based psychosomatic medicine and psychotherapy is perhaps the most likely result. But it would penetrate deeper than that. For the first time man would possess a fully authentic image of himself as a unitary being, and this would certainly have important social consequences. I outlined some aspects of this in *The Next Development in Man,* and shall not attempt any further analysis here. I will only add that it places an unusual strain on the mind to be aware on the one hand of the arsenal of bombs, and on the other of this unprecedented path leading, it would seem, toward an at least partially redeemed humanity. These apparently sharp alternatives of evil and good are a consequence of the fact that we are now at a threshold where we can see two vistas: the past trend toward collapse, and a possible future trend toward liberation from some at least of the causes of the past. In 1972 my thought can carry this no further.

5. With somewhat greater confidence in my judgment I put the following negative on record. I find it difficult to believe that during the next three decades the human mind will become fit to understand its own paranormal faculties. I mean by this term those capacities which I believe exist, but which it may be misleading to describe as telepathy, the experience of apparently nonrational coincidences, clairvoyant powers, and so on. It seems to me that many decades of research and interpretation, facilitated by the unification considered under 4 above, may be necessary before any definitive clarification of this challenging realm can be reached.

The matters raised under 3 and 4 above are closely related. I cannot deny the awareness of great possibilities for man which led me to write *The Next Development in Man* thirty years ago, and which remains with me now. But one man's vision, even if it contains some elements of a lasting truth, must be partial, even lacking in balance. Yet I cannot fail to end on a positive note. The United States long lived by the *American Dream*, but this has now faded. The prospect of a unitary dispersal of the body/mind dialectic curse, with all ensuing benefits, I will call the *Human Dream* of the late twentieth century.

Notes

These notes serve several purposes. (1) In some cases, before the numbered notes to each chapter, references are given to publications which I regard as valuable on the subject of the chapter. (2) References are given, supporting the argument of the text. (3) The argument is developed in more detail. (4) I have used these notes to aid the advance of clarity and unity in science and elsewhere by calling attention to relatively new points or to recent publications, which in my view have received insufficient attention.

Chapter 1

1. My autobiographical volume *Focus and Diversions* gives more detail on some of the themes of this chapter.
2. I consider that the history of an idea is indispensable to an understanding of its present meaning, and in my philosophical reading I have sometimes come across strangely neglected figures. For example Alexander Bryan Johnson's *A Treatise on Language* (1836; California, 1959) is of great interest and expressed some of the ideas later enunciated by Mach, the Vienna Circle, and Wittgenstein. Yet I believe that no British writer on Wittgenstein or on semantics has mentioned Johnson. This is a striking case of the widespread neglect of the history of ideas. For a one page summary of Johnson's ideas on the role of the language see L.L.W., "Early Semantics," *Philosophy*, July 1960, p. 272. Fortunately a conference on Johnson's life and work was held in 1967 in Utica, New York, so he is not forgotten in the U.S.A.
3. *The Unitary Principle in Physics and Biology*, and *Accent on Form.*
4. In *The Next Development in Man* (hereafter referred to as *NDM*) "Unitary Man" was defined as a more harmonious type marked by his conviction of a universal formative process, here designated more precisely as a universal morphic process in 3D. See chapter 3, and note 4 on "morphic."
5. Albert Einstein, *Ideas and Opinions* (New York: Crown Publishers, 1954) , p. 48.
6. In *NDM* the "European dissociation" was defined as the separation of deliberate from spontaneous behavior. It has had many different forms—religious, moral, financial, and so on.
7. G. F. Barbour, *The Life of Alexander Whyte* (London, 1923), p. 532.
8. *Focus and Diversions*, p. 19.
9. Henry Drummond (1851 1897, Scottish Evangelical writer). See his *Natural Law in the Spiritual World* (1883), p. 35. It seems that my memory has improved on his text. Drummond's influence, was greater than is generally known. For example, my cousin George Brown Barbour (later professor of geology and until recently dean of the College of Arts and Sciences at the University of Cincinnati) was working in

China (1923 34) with Pierre Teilhard de Chardin, who at first knew little of earlier attempts to unite scientific knowledge and inner experience. Barbour conveyed to Teilhard during many intimate talks Drummond's ideas on the unity of all realms of experience, recommending to him some of Drummond's books. Mrs. George (Dorothy) Barbour also discussed with Teilhard Drummond's sense of a pervasive unity. This was all prior to Teilhard's development of his own doctrine. (Why did he not refer to Drummond?) The fact that Drummond's ideas had influenced Teilhard as well as myself has only just (May 1972) come to my knowledge. But Teilhard's and my view of the underlying unity diverge radically.

10. I shall consider this great power in chapter 3. It rests fundamentally on the greater logical scope of asymmetrical relations over the corresponding symmetrical relations.

Chapter 2

I know of no book which adequately covers the new situation and role of science. But see M. Polanyi (note 2, below), 7. R. Ravetz, *Scientific Knowledge and Its Social Problems* (1971), and particularly good: T. Roszak, "The Scientific World View," *Nation*, March 22, 1972, a review of Ravetz's book.

1. Science can no longer be regarded as "the pursuit of objective knowledge or facts by empirical methods," since the methods used determine the facts observed. Thus the concept of "demonstrable facts" is now seen to be nugatory. Even in physics a single expression is seldom tested in isolation, but rather the compatibility of an entire section of a theory with an extended set of observations. It is futile to try to restore the traditional concept of scientific objectivity. What is now necessary is to discover what property should replace the old "objectivity" in the new scientific situation. All the sciences are now in the realm of multifactor systems and asymmetrical rela-tions, for which no general methodology is yet available.
2. M. Polanyi, *Personal Knowledge* (London, 1958).
3. Henri Poincare, *Dernieres Pensees* (1913), p. 20. A science which fails to cover subjective experience is incomplete.
4. There is no comprehensive unified theory of all known particles and forces. I know of no book on the philosophy of physics since 1950 which I can recommend. Anyone wishing to learn something of the subtlety and complexity of important advances in physics should read N. R. Hanson's excellent *The Concept of the Positron: A Philosophical Analysis* (1963), the best study of an important recent advance in quantum theoretical physics known to me. Decades spent searching for a unified field equation have proved fruitless. It may now be time to consider instead a unifying inequality. On physics see also chapter 4, note 4b.
5. Biological theory is immature. There are only two approaches to an authentic biological theory: (a) evolutionary theory and (h) molecular biology.

a. Even after a century there is no theory of the origin of mutations covering changes, that is, the precise disturbances of the genome (chromosomes, and so on) by radiation or otherwise; and the development from these disturbances of stable mutations (see note 8, below). Moreover, I believe there is no description of the total conditions affecting any major evolutionary advance. The neo Darwinian synthetic theory may express necessary conditions, but not sufficient conditions for evolutionary changes. Little is known about favorable mutations and next to nothing about the evolution of speaking man. Further, the mathematical theory of evolutionary rates based on information theoretical methods cannot be valid, since the internal constraints (hierarchical structuring, and so on) in organisms have not been identified and, as Quastler showed, no numerical results can be reliable until such constraints are understood and allowed for. Errors of the order of 10^6 and more can easily occur.

For fundamental problems in biology a comprehensive theory of morphogenesis is necessary, covering the entire ontogenetic process including repair processes, for example, in man from the fertilized ovum to the operation of due central nervous system and brain, including the development of memory and of language. A complete biological theory must also cover the emergence and evolution of life and the hereditary aspects of behavior, as expressions of a single set of basic principles.

b. Molecular biology is of great importance and will doubtless achieve further successes. But until molecular biology, or the quantum biochemistry of complex structures, is extended to provide an appropriately simple representation of hierarchically structured system., each as a protein molecule, advance will he hampered. A desirable step would he the identification of global variables characterizing each level, and of their relations to the local biochemical variables.

6. It is shown here that even men of the caliber of Maxwell and Einstein can make what prove to be serious mistakes. Frequently during the history of exact science leading figures, usually expressing the tenets of an orthodox school, have made egregious blunders. The most striking of these often take the form of assuming that the knowledge of some region of science is complete and that its basic concepts and laws are already known. The quotation given in the text shows that in 1875 Clerk Maxwell believed that the concepts of classical mechanics held final truth. Around 1930 to 1935 many theoretical physicists considered that the quantum mechanics of that time was the definitive physical theory. More recently one school has suggested that molecular biology covers all biology.

Today science needs professional watchdogs, showing up orthodoxies when they become arrogant and deny a proper hearing to new ideas. But few individuals are qualified to fill this role of watchdog. They must be influential, so that their bark is heard; they must know something about several sciences; they must be courageous enough to attack established cliques; and, most important of all, they must have a sense of what is in the air and where new ideas are likely to be necessary. Arthur Koestler meets these conditions, and several of his books have been of great benefit to science by challenging exhausted orthodoxies. This valuable contribution to science may be summarized thus:

Orthodoxy	**Book by Arthur Koestler**
1. Marxist-Leninist	*Darkness at Noon; The Yogi and the Commissar*
2. Naïve inductive view of science	*The Sleepwalkers*
3. Behaviorism	*Insight and Outlook; Art of Creation; Ghost in the Machine*
4. Synthetic neo-Darwinian theory	*Ghost in the Machine; Case of the Mid wife Toad*
5. Zoomorphic Image of Man	Two essays in *Drinkers of Infinity*

In these cases I accept Koestler's criticisms.

Until around 1965. I believed that the rather high standards of keenness, precision, and objectivity characteristic of theoretical physicists would be shown also by biologists interested in theory. But there is no community of theoretical biologists continually discussing and criticizing each other's ideas, though C. H. Waddington has done something to create such a group.

7. See note 6.
8. It is widely held that "mutations are random," by which is meant "not correlated with adaptive properties." This assertion is today ambiguous and misleading, as the term "adaptive" acquires a wider meaning when attention is paid, as is necessary, to internal adaptation. All organisms are now known to be highly structured hierarchical systems and any properly functioning part most adequately conform to its local structural environment. Thus a closer analysis of the term "random mutations" is now necessary, taking into account the new knowledge about biostructures. (A first attempt in this direction was made in L.L.W., *Internal Factors in Evolution* [1965], p. 100.)

First, it is necessary to separate the original unstable and random *changes* in the chromosomes, or in a gene (often caused by radiation), from any resulting mutations which are stable enough to serve as hereditary determinants.

Second, the initial random structural change can lead to two results:

a. The changed structure is modified (1) so as to bring it into stable physical equilibrium with its local environment, and (2) so that it sufficiently conforms to its local environment to be functionally stable (in replication, and so on). This modification results from forces operative at the several molecular levels. In this case the random change is converted into an *internally adapted mutation*, determining a phenotype subject to the standard Darwinian environmental selection. *Such internally adapted (not random) mutations constitute the sole basis of evolutionary advance.*

Or: b. If the random change is too great, or too alien to the local structural environment, to be thus adaptively modified, it persists as a deleterious or lethal mutation, distorting the growth and lessening the survival value of the system.

The above cursory analysis neglects the distinction between the various structural levels. A mutation which is internally adapted at one level may not be so at another. Thus a particular mutation may be favorable in one respect (that is, at one level) and deleterious in another respect (that is, at another level), as is known to be the case.
9. James Clerk Maxwell, "On the Dynamical Evidence of the Molecular Constitution of Bodies," *Nature*, XI (March 4, 1875), 357. Also, *Collected Scientific Papers*, II, 418.
10. Albert Einstein, "On the Electrodynamics of Moving Bodies," *Ann. d. Phys.*, 17, 1905. Translated in *The Principle of Relativity* (collection, 1923), p. 39.
11. See G. J. Whitrow, *The Natural Philosophy of Time* (1961), p. 29, giving references to Aristotle's *Physica*.
12. A. S. Eddington, *The Nature o/ the Physical World* (1928), p. 80. Eddington introduced the term "time's arrow" to express a directed or one way property which has no equivalent in space.
13. All processes which lead to a more regular or symmetrical state (that is, morphic processes: see chapter III) display time's arrow; indeed two major classes of physical processes, neglected by Eddington, did so in the physics of 1927: morphic processes and retarded potentials in electromagnetic theory.

There is a point here of interest to students of simultaneity. In 1927 Eddington and I were both interested in time's arrow. Eddington's Gifford Lectures (see note 12) were given in 1927, and published in 1928. During 1927 I published *Archimedes, or the Future of Physics*, which was mainly concerned with "irreversible processes" (in some respects a misleading term—*one way* is better), and on pages 95–96 gave the first list of these known to me, which included "processes dependent on retarded potentials." I sent a copy to Eddington, knowing of his interest in the "direction of time" from his *Internal Constitution of the Stars* (1926): If he had read it, he could have corrected his mistake in proof by including retarded potentials. There are today (1972) five recognized classes of processes containing tune's arrow: morphic processes, retarded potentials, entropy increase, the expansion of the universe; and subjective experience. See chapter III,, note 3.

14. A, L. Lehninger, *Bioenergetics* (New York, 1965), p 4.
15. Every student of the history of science knows the remarkable mistakes made by some of, the greatest names, often with sublime self-assurance and air of authority. Several recent controversies have displayed a shocking degree of carelessness and lack of the objectivity scientists usually claim. Examples are: the nature/nurture controversy, particularly in rotation to the potential intelligence of blacks in the United States, and the pollution controversy. An instructive example of how mistaken conclusions can become part of an assumed tradition of scientific truth is given by H. Elias, "Three dimensional Structures Identified from Single Sections," *Science*, 174 (Dec. 3, 1971) 993–1000.
16. Anyone wishing to .explore the potentialities of his own mind will find M. Drury, *The Inward Sea* (1972), a delightful study of the "riches within us," using an introspective approach.
17. The mechanical view based on "motions of matter" first proved inadequate around 1880 to 1900, when it was found impossible to provide a mechanical model for Maxwell's field quantities. Since then every fundamental advance has further lessened the scope of classical mechanical theory. Today fundamental physical theory uses structures, groups, probabilities of observables, and so on, but offers no clear view of the basic character of the physical universe. This situation is unlikely to persist for long.

Chapter 3

1. Some of the recent discussions on chance and determinism have, in my view, been premature and, misleading. The authors usually assume that enough is already knows about fundamental principles to provide a reliable approach to this question. This is wrong; we know next to nothing about fundamental laws, no science being yet unified. It is possible that the statistical character of quantum mechanics, as now understood, may not survive in a unified theory. Einstein may prove to have been partly right. God may not play dice, because he is interested in order, not in numbers.
2. These tendencies or one way processes (defined as leading toward a terminus) are finalistic in the formal sense of moving toward a final state, but they should not be regarded as involving "final causes," their context being non Aristotelian. They require the use of global variables. "Finalistic" is a general property; "teleonomic" is used by some for organic finalistic processes.

   The use of *global geometrical diminants* (signless quantities representing a deformation from symmetry which decreases in a movement toward symmetry would give precise expression to the "holistic" or "system" property vaguely designated by such terms as "fields" (in biology), "integrative wholes," "organicism," and "the unity of complex systems." Thus a whole can (loosely) be said to be more than the sum of its parts, if and only if the whole obeys global laws. But it is not yet known whether invariant local atomic variables or diminant global variables are more fundamental, that is, provide the better basis for a unified theory. (See note 22 to "The Structural hierarchy in Organisms" in *Unity through Diversity*, A Festschrift for Ludwig von Bertalanffy, ed. William Gray and Nicholas D. Rizzo, Part I: (N.Y. Gordon & Breach, 1973). If global dominants are necessary to a unified theory, then *all apparent constants* in current theory (masses, charges, spine, and so on) are actually diminants whose rate of decrease is negligible when certain (quantized) values are passed through.

3. The confusions regarding entropy and ordering processes are instructive, and provide opportunities for entertainment and intellectual exercise. Leftover, controversy aides science. Here are two examples, further to those in the text:

A. A few years ago a distinguished theoretical physicist, whom I regarded as wise, said something over the lunch table that provoked me to protest, "That would not be true of processes in which order increases," to which he replied, "There are very few of those." After I had listed six or even seven classes of morphic processes on the back of a menu. he said, "Sorry, I was wrong." (An extensive list of morphic processes is given in note 4 below.) I learned much from that episode, but I lost an admired father figure. Probably he had been thinking only of conventional closed systems. As it was a private occasion I do not give his name.

B. Here is a more recent, subtle, and instructive, example. Professor H. J. Morowitz of the Department of Molecular Biophysics and Biochemistry at Yale, in an interesting paper on "Biology as a Cosmological Science" (read at the New School, New York City, in October 1971, and published in *Trends in Modern Thought* [New York, 1972), includes this phrase: ". . . since its introduction over a century ago no evidence has come forth to challenge the second law of thermodynamics:" This is an old cliché but, I submit, misleading to the innocent:

a. It is ambiguous. Entropy meant different things at different times. Statistical mechanics and information theory have increased the precision of the second law. To which formulation of the law is he referring?

b. "No evidence." This is curious. For a quirk survey of the physical universe suggests antientropic or overall ordering tendencies at many levels, and entropic, tendencies toward disorder only locally. Most cosmologists today assume that the universe has passed from an early undifferentiated state, such as a primeval gas or a fireball, toward the highly differentiated state containing complex atoms, molecules, minerals, solar systems, nebulae, and so on, which we observe today (though great time lags are involved). So scientific common sense (1972 version) suggests that the traditional entropy dogma, in its verbal, vague, and unrestricted formulation, is misleading (perhaps because in a fundamental analysis closed systems are very rare).

c. Moreover, during the last few years some of those concerned with statistical mechanics and information theory have come to believe that the second law is only fundamentally valid where "preferred states or configurations," "molecular structures," or "micro information" are absent. This means either that in a fundamental analysis the second law never applies exactly to the real world in which structures are everywhere being formed and that it is only an approximation valid at that level of analysis in which structure can be neglected, or that this entropy concept must be extended and given a new meaning.

Whatever one's interpretation of the position, caution is clearly necessary. In my view the quoted phrase is ambiguous and misleading, for it neglects the cosmological trend from an undifferentiated toward a differentiated state which can hardly be questioned.

For an up to date and reliable discussion of the status of the second law in the statistical mechanical theory of nonequilibrium states see P. T. Landsberg's review of Glansdorff and Prigogine, *Thermodynamic Theory of Structure, Stability, and Fluctuations* (1971), in Nature, 238 (July 28, 1972), 229-230. I regard Landsberg as one of the best writers today on the precise range of validity of thermodynamic concepts and principles in statistical mechanics.

Join the radio astronomer and consider how he sees the Crab Nebula, as a fantastically delicate and complex filigree structure with countless filaments forming shells. This is believed to have arisen from an earlier, less structured state under the influence of magnetic fields, for these tend to establish en unmistakable geometrical order, immediately observable in 3D, by aligning magnetic entities (spinning charged particles and magnets) into orderly arrangements. It is a far cry from this unmistakable 3D order to the abstract theoretical constructs in phase space of statistical mechanics.

Turn from the cosmic to the minute. The application of the entropy principle to the first formation of a minute crystal nucleus, indeed to crystals in general, has not yet been fully clarified. All we know for certain is that the loose correlation of higher entropy with higher "disorder" (of what kind: dynamical, kinematic, or geometrical?) is treacherous, being often either ambiguous or false. If as a last resort we turn to Willard Gibbs and his early view of entropy as meaning "mixed upness," we quickly discover how limited is its scope. Nature, as we actually see it, is far from being "mixed up." Even reel fluids display transient regions of higher order where a fluctuation has occurred. The glib entropy dogma encourages us to treat fluctuations as merely secondary effects, whereas they are a necessary micro aspect of macro equilibrium states.

The truth is that in the real universe out there in 3D, and not in our theory clogged minds, morphic processes predominate. A thermally minded physicist may say, "That's only because the temperature is low enough." Exactly. If it were high enough everywhere the universe would not be the one which exists, it would he without structure, form, life, or mind, and no Crab Nebula could ever be formed. There are certainly local entropic effects, but the universe as a whole, on most of its levels, is such that structure can develop.

The best way to clarify our minds on entropy, and to avoid the difficulties in applying statistical mechanics and information theory to fundamental problems (problems of definition, selection of representations, micro/macro variables, and so on), is provisionally to forget the concepts used in these theories, and first to concentrate on Eddington's brilliant concept time's arrow. This is a wonderful aid to clarity because it is vivid and draws our attention to the immediately given (3D, S). In the following, I use time's arrow to present my own interpretation of the present (1972) position of theory and fact in this realm. There exist today arrows of very different status:

A. *The Morphic Arrow* (Layzer [1971, see below] calls this the "Historical Arrow") is the most extensive and directly observed, and therefore probably fundamental in a unified theory. It is defined by morphic processes, that is, increasing symmetry or order in structure generating processes in certain open systems. Apparent reversals of these processes actually follow a different path (see chapter 5, note 7). *This and E below are the only arrows that ore immediately observed*, not merely properties of a provisional nonunified theory.

B. *The Electromagnetic Arrow* expressed in the fact that *retarded potentials* are more widely applicable than advanced ones. This arrow is not directly observed but is dependent on EM Theory, which cannot survive unchanged in a unified theory. Radioactivity may be an expression of this or a closely similar arrow.

C. *The Thermodynamic Arrow* of increasing *entropy* in closed systems without structure, which are rare. This depends on the use of entropy theory, and, in the light of recent work, may prove to possess little or no fundamental significance, as preferred states and structures are always present.

D. *The Cosmic Arrow* present in the *expansion of the universe*, which rests not on direct observations, but on a theoretical interpretation which is speculative and may prove unnecessary.

E. *The Subjective Arrow* determined by the immediately experienced sense of *later than*. This does not depend on theory.

Note that *The Morphic Arrow* and *The Subjective Arrow* are directly recognized in (3D, S), whereas The *Electromagnetic Arrow*, *The Thermodynamic Arrow*, and *the Cosmic Arrow* all depend on the validity of certain current theories.

The ultimate source of the difficulties surrounding entropy lies in the fact that physicists have for long used symmetrical relations (measured quantities, reversible processes, and so on), whereas in the realm of entropy, morphic processes, retarded potentials, the expanding universe, and the experience of succession, asymmetrical relations are required.

I recommend (a) on entropy, cosmic evolution, and thermodynamics, D. Layzer, "Cosmic Evolution and Thermodynamic Irreversibility;" *Pure and Applied Chemistry*, 22 (1970), 457, and "The Arrow of Time," *Vistas in Astronomy*, ed. by Beer (1971), p. 279. Any clarity I now possess results from the fact that the advance of thermodynamic and information theory (for example, Layzer) has now (1971/2) converged to meet my own ideas on asymmetrical relations. I wish to express my indebtedness to David Layzer for offprints and correspondence, though his terminology and classification of the various arrows of time is not identical with mine. (b) On entropy in single complex molecules, McClare's work (see chapter 4, note 4b), which I regard as important.

Any failure (in the particular representation used) of invariant degrees of freedom determining discrete accessible and enumerable states of a system demands reconsideration of the concept of entropy and of the second law. Such failure may be caused by interactions leading to a structure stabilized by constraints, or in a future theory by the substitution of a primary continuous process for quantized states.

4. The following are examples of morphic processes:
Processes of the formation, from prior less ordered or less symmetrical states, of the following forms and structures:

Complex nuclei, atoms, molecules, polymers.
Crystal nuclei, snowflakes, macro crystals, liquid crystals.
Ordered phases, structures, and groups in liquids.
Soap bubbles.
Ordered states in solid alloys.
Systems of high order at normal or low temperatures:
lasers, masers, superconductors.
Clusters of particle tracks, rainbows, vibration patterns, tornadoes,
shaped clouds, moon craters and ridges, solar systems, star groupings,
structured galaxies.
The preorganic and organic structures marking stages in the *morphic continuity* of the evolutionary process leading to *Homo sapiens*: for example, heterogeneous polymers, catalytic chains, ending structures, and the earliest pulsating hierarchies composed of DNA, RNA, proteins, etc. All self assembling and replicating systems. Enzymes, multienzyme complexes. Fibers, platelets, membranes. Mitochondria, other organelles, mitotic structures, viruses. Cells, tissues, organs,

adult organisms, and their differentiated parts. Spatially ordered colonies. Any transient or lasting structures in the brain underlying aspects of mental processes: memory, reason, etc. All man-made forms.

Most of my writings since around 1930 have been concerned with the "formative" processes by which visible 3D forma are generated. After waiting for many years for someone else to recognize the necessity for an improved name, by 1966 I considered that the time was ripe. I therefore defined and used the term "morphic" in a sequence of papers written from that date onwards. There is evidence that the term is timely. For example, a mathematician has recently called attention to the fact that neither in physics nor in biology has morphogenesis yet received an adequate mathematical treatment. See R. Thom, "Topological Models in Biology," in Waddington's *Towards a Theoretical Biology*, III, 89.

5. This was observed independently and nearly simultaneously by A. Sellerio of Palermo, B. Renaud of Paris, and myself around 1935. For references see *The Unitary Principle in Physics and Biology* (New York: Henry Holt, 1949), pp. xi, 19. It is likely that so obvious a parallel was also noted by others, earlier, though few have put the main emphasis on a tendency toward symmetry. See *Unitary Principle*, pp. 14–19. The parallel of maximum symmetry and maximum entropy was implicit in Gibb's statistical mechanics.

6. From 1950 onwards, and particularly around 1966–67, a few frontier minds began to realize the importance of hierarchical structures in exact science; by 1972 there exists already an extensive literature on the subject. (This is an example of the simultaneity discussed in chapter VII.) For a historical survey of the concept from early times on, see my "The Structural Hierarchy in Organisms," note 2 above, and for an exhaustive bibliography up to 1968 see Donna Wilson's excellent paper, "Forms of Hierarchy: A Selected Bibliography" in *Hierarchical Structures*, ed. by Whyte, Wilson, Wilson (1969). The theme of hierarchy in the psychological and social sciences is treated in *Beyond Reductionism*, ed. by Koestler and Smythies (1969), and in biology in the four volumes of *Towards a Theoretical Biology*, ed. by Waddington (1968 72). See also my Korzybski Memorial Lecture (1969), "On the Frontiers of Science: This Hierarchical Universe," *Gen. Semantics Bull.*, No. 36, pp. 7–14.

From 1970 on, many scientists became interested in hierarchies. As an example of these I recommend J. Bronowski, "New Concepts in the Evolution of Complexity: Stratified Stability and Unbounded Plane," *Zygon* V. 1 (March 1970), 18, though, in my view, he exaggerates, as is still common, the role of chance, and inadequately emphasizes the part played by order generating (morphic) processes. Nonetheless his paper is helpful, particularly in its stress on the fact that the second law, as normally interpreted, presupposes the absence of preferred configurations.

Arthur Koestler (ref, above) has introduced the useful term *holon* for any unit which (a) contains parts, and (b) is also a part of a larger unit, that is, for all units in hierarchies excluding any largest and smallest units.

7. No work is known to me which discusses the general status and importance of asymmetrical relations in science, the nearest being (after Russell's and others' early studies on time) works restricted to time and neglecting the theory of asymmetrical relations, such as H. Reichenbach, *The Direction of Time* (1956, now outdated), and G. J. Whitrow, *The Natural Philosophy of Time* (1961, good on the ground covered, but neglecting the TCP Theorem, and outdated by recent work on entropy). Apparently no one has yet used the logical theory of asymmetrical relations to throw light on empirically identified asymmetrical relations (less than, later than, hierarchical relations, broken symmetries, and so on). This I find baffling. Russell discussed

asymmetrical relations from 1914 on. The contrast symmetrical/ asymmetrical relations is the most profound in the logic of science, and physicists have been troubled by entropy for a century. Why has no mathematician, theoretical physicist or philosopher of science taken up this timely, fertile, and intellectually intriguing issue? The importance of asymmetrical relations is obvious, for they can contain the corresponding symmetrical relations as a special case. (All the ideas in this book which are either new or specially emphasized are the expression of asymmetrical relations.)

When in science one replaces symmetrical relations by asymmetrical relations one enters a new, almost totally strange, and highly fertile realm. Indeed in this novel realm only one model can lead (or perhaps mislead) us:

$$S_{t2} > S_{t1}$$

in a closed system neglecting structure (S = entropy). I say "mislead," because the orthodox view is that the essence of this principle lies in its statistical character, and this may not be what is required in a general theory (see note 3, and chapter 4, note 4b).

For long I have waited for signs of pure or applied mathematics not only realizing that inequalities are logically prior to and more general than equalities, but paying more attention to their scientific applications. But alas, no one has yet developed a general theory of asymmetrical relations in science.

However, in a neighboring realm an important signal has appeared. George Spencer Brown, in his work of genius, *Laws of Form* (London, 1969; New York, 1972 73), has developed a new symbolic calculus which received Russell's blessing, based not on number, class or set (symmetrical relations), *but on the operation of drawing a distinction*, which, as he uses it, is an asymmetrical relation. For Spencer Brown *mathematics* begins with the drawing of a distinction of an asymmetrical character; I believe *physics* should begin with the distinctions "less than" and "later than." I recommend to all interested in the frontiers of the intellect the introduction and notes to *Laws of Form*, though many will find his symbolic calculus too new and strange to be easy to absorb. If Spencer Brown's calculus finds further empirical applications—he believes it already has in the design of electrical circuits—this book may prove to be one of the most important in its field of the late twentieth century. See my review of *Laws of Form* in *B.J. Phil. Sci.* 23 (1973): 291 294.

In using the complex symbol (3D, S) to represent the immediately observed relations of distance and succession, I have here neglected for the sake of brevity and clarity a point of importance in a fundamental analysis. In actual fact the immediately observed spatial relation is not the absolute equality of two distances, but the observed asymmetrical relation "less than" between two distances, or its absence.

An unfortunate blunder was made when Minkowski and Einstein encouraged scientists and others to think that the Special Theory of Relativity implied that temporal relations were fundamentally and in general similar to spatial relations. In 1908 H. Minkowski published the famous words: "Henceforth space by itself, and time by itself are doomed to fall away into mere shadows, and only a kind of union of the two will preserve an independent reality." Later Einstein wrote in his Relativity: The Special and General Theory (1921): "There is no more commonplace statement than that the world is a four dimensional space time continuum."

If is my duty to point out that these assertions, taken literally or out of context, are misleading. Whose "world"? For whom or in what theories do space and time "fall away into mere shadows"? These assertions are not valid of the world of human experience, nor for the biologist dealing with life. If we are concerned with these realms we have to unlearn fifty years of specialized thinking, which has been interpreted too widely.

The assertions made by Minkowski and Einstein do not apply to any "real world," if that has meaning, but only to a particular type of theoretical representation of a narrow class of processes using external metrical coordinates and treating c (the velocity of light) as fundamental. They assert a property of a particular type of theory, not of other theories, and not of the directly experienced world of (3D, S).

4D space-time is a theoretical construct which is most useful, I suggest, only for relatively simple systems in which certain physical interactions are neglected. When complex systems are treated or fundamental interactions covered, theory becomes embarrassingly abstract and it may be found better to use internal coordinates to represent the one-way process of a complex system relaxing toward equilibrium. In any case the 4D space-time continuum is a theoretical construct whose range of applicability is not known; moreover, most quantum theorists rely on a quantized discontinuum. A full critique of the 4D space time concept leads into the realm of relativistic quantum mechanics, which, in its present form, is unlikely to prove definitive.

It is probable that had Minkowski or Einstein been challenged on this around 1930 (actually Minkowski died in 1909), they would have accepted a restriction on the range of validity of these assertions. Einstein knew by 1929 that his Special Theory might not be appropriate for representing complex systems such as a protein molecule. During the winter of 1928–29 I had two talks with him, and my main hope was to obtain his blessing for my preoccupation with asymmetrical relations and "irreversible" or "one-way processes" in certain systems where, owing to the presence of ordered structure, classical entropy might not be relevant. On my pressing him, to my great relief he said—as I remember it—"Yes, of course for describing biological or complex molecules, relaxing towards states of minimum potential energy, the relation of succession may be necessary, and the methods of the Special Theory may not be appropriate." My memory must be correct, for I remember my pleasure at his remarkable intellectual generosity and readiness to encourage a younger mind struggling to obtain clarity in a realm different from that of his own main interests. I came away greatly reassured.

The greater power of asymmetrical relations over symmetrical relations suggested to me in 1929 (see *Focus and Diversions*, p. 171) that the most comprehensive law of ordering and therefore of physics might have the form:

$$At_2 < At_1 \text{ (A = some 3D asymmetry)}$$

in any system isolable for the purposes of scientific representation.

8. See T. Roszak, *The Making of a Counter Culture* (Garden City, N.Y.: Doubleday, 1969), p. 254; also A. Maslow, *The Psychology of Science* (New York: Harper & Row, 1966), p. 16.

Chapter 4

1. Bertrand Russell, *Principles of Social Reconstruction* (London: Allen & Cowin, 1916), p. 245.
2. Letter of December 19, 1966, to Mrs. R. Heywood, who kindly gave me permission to use it. "Numinous" is Rudolf Otto's term for "sacred" or "holy."
3. For example, N. Hartmann, *Philosophische Grundfragen der Biologie*, 1912. I have written on a hierarchy of formative processes since 1944 (see index to *NDM*). This idea has been discussed by many since, the late 1960s. (See, for example,

Bronowski, chapter 3, note 6 above.) The world view is in my opinion only slightly ahead of the scientific orthodoxy's identification of particular applications.

4. This note on the unification of knowledge is divided into four parts: (4a), philosophy; (4b), physics; (4c), psychology; (4d), general systems. Biology is treated in chapter V and notes.

4a. The world view is "transcendental" in a similar sense to that of Kantianism, (i) as expressing the necessary conditions of human experience as determined by the constitution of the human mind; and (ii) as transcending the contingent particularity of experience.

Indeed, the world view can in some respects be regarded as a *Kantian Transcendental Aesthetic of the Late Twentieth Century*, a doctrine of the forms of our perceptions of space and time, interpreted as relational in (3D, S). I cannot here examine the relations of the world view to traditional philosophies. On the relation of the world view to remain psychological and religious attitudes see below, 4c, on "levels."

4b. It is unusual, though far from unprecedented, to attempt to predict the optimal path of advance of fundamental physical theory. I expect that many of the difficulties and paradoxes of current theory will vanish in a theory based on the world view and (3D, S).

Global relations, including object and measure, are macroscopic and immediate, involving (3D, S). The most powerful physical theory must probably use global variables based on (3D, S) and representing asymmetrical relations. (See chapter 3, note 7, above.) Any other type of theory, for example, treating symmetrical relations and local variables as primary, is necessarily more restricted in its scope.

Electromagnetic (EM) effects possess a unique status, and an improved theory might reasonably grow out of a reinterpretation of quantum EM. As a first step I have adopted the heuristic hypothesis that a skew-deformed tetrahedron relaxing toward the regular form can provide a *geometrical* model of quantum EM, in contrast to the *mechanical* models of classical EM sought for without success by Maxwell and others. (See *Studium Generale*, 23 [1970], 525.) Observations of λ in diffraction involve at least four points (source, sink, and two diffracting centers), and these suffice to determine a tetrahedron.

A unified theory must *inter alia* derive electron theory (e^2), quantum mechanics (h), and relativity theory (c), from a single principle, as Einstein saw in 1909 (*Phys. Zeit.*, 10 [1909], 192 93). This implies that in a unified theory c cannot represent the linear velocity of a harmonic EM wave, but most be reinterpreted. Anyhow, in a fundamental theory the velocity of a pure harmonic wave cannot be measured, as the markers or interactions required to signal departure and arrival destroy its pure harmonic form.

These observations arise from an application of the one-way processes of the world view to physics, since constants such as e^2, h, and c cannot play a primary role in a one-way process, but must represent most nearly stationary components of such a process. The requirement of the solution of a problem outstanding since 1909 is a severe test for a philosophical world view.

This argument suggests that a continuous process may underlie the apparent quantization by h, as has been argued by M. Sachs (*B.J. Phil. Sci.*, 15 [1964], 213, and 21 [1970], 359) from a more abstract point of view. Einstein also hoped that a unified field theory would lead to a continuous process accounting for the quantization of the quantum mechanical equations. At one time he considered using overdetermination, that is, employing more field equations than unknowns, to obtain special (quantized) values, but without success.

Though it lies in a field in which I am not competent, I recommend to those interested the work of C. S. Smith of M.I.T. on the role of surface energy in determining the equilibrium forma of polycrystals (Metallurgical Review, 1963, and elsewhere).

I now pass to a new development in entropy theory. In an important series of papers (*Journal of Theoretical Biology*, 30 [1971], 1 34; 35 [1972], 233 46; and 35 [1972], 569 95, the first of these being the most relevant to entropy), Colin McClare makes what I regard as a basic contribution to thermodynamic theory. (The novelty of McClare's work has been challenged, but his emphasis on the importance of separating two types of energy exchange is likely to prove fertile.)

McClare argues: (i) that the second law is not necessarily statistical but can be applied to single systems, such as a complex molecule, inorganic or organic; (ii) that energy exchanges are separable into two sharply distinguished classes: (a) rapid quantum mechanical resonant transfer of energy in single molecules resulting in "stored energy," and (b) the ordinary very much slower thermal exchanges by collisions, etc. The ratio of the rates of the two processes can vary greatly, say from 102 to 1010; (iii) that processes of class (a), in the cases considered, involve a change of enthalpy, and provide an explanation of energy effects in biomolecules often wrongly ascribed to classical thermal machines. Thus (a) provides a model for the transformation and storing of energy in single ATP molecules, applicable to muscle action, active transport, oxidative phosphorylization, etc.

4c. The world view points to an urgent task in psychological theory: representation of the *functional levels of the mind*, in terms of brain physiology, or of emotional and cognitive processes, or better: uniting both, and linking them with the underlying neuroglandular physiological levels of the animal organism. This should not separate body and mind, but treat the mental levels as an extension of the levels of the structural hierarchy of the organism. Such a theory of mental levels should pass from the most specific, focused, varied, and usually conscious processes of perception through various levels to the basic, most general, unified, unitive, and usually unconscious level constituting the basis of all thinking. A theory of mental levels would explain induction (Hegel's "cunning of reason"), imagination, and the establishment of values (see chapter V on "vital surplus"). This doctrine of mental levels should cover all exoteric and esoteric "religious" experiences, including sudden "conversion," and would interpret Yoga and many of the Eastern religions as providing methods designed to heighten awareness of the basic unifying level of the mind, thus easing control of other "higher" levels. The West has not yet recognized, as the East has for long, the crucial role played by a basic or ground mental level in all human experience. The conception of graded levels in the human mind is a step toward greater clarity on many matters more stressed in Eastern than in current Western thought, for example, levels of truth, different aspects of personality and of the inner life, the eternal Self, the ecstasy of unity, the experience of grace, and so on. (Such a theory would not correspond closely to Freud's Superego, Ego, and Id, which is a highly complex system and which erroneously presupposes a basic libidinal level.) I have not found any useful approach to such a theory of functional mental levels.

4d. We pass now to general theory of physical systems. The world view implies that open systems losing energy may possess a fundamental theoretical importance. I therefore wish to call the attention of mathematicians, scientists, engineers, or systems theorists to a development which, though still at a very early stage, may prove of importance. A group of French engineers and systems theorists (calling themselves SYSTEMA, 30 rue Croix Basset, 92, Sevres, France), believing that differential equations of the traditional types may not be appropriate for representing complex open systems, are endeavoring to develop instead an arithmetical calculus using integers, based on "arithmetical relators" employing recursive (iterative) relations between integers. They are concerned with open systems, dynamic processes, a calculus using integers, and systems of finite complexity, and they employ three interdependent aspects of any description: "concrete reality," "action," and "information," appropriately defined. It is at too early a stage for me to express any view regarding its fertility.

5. The history of scientific and philosophical thought throughout this century shows a trend toward relational methods. But this has not yet culminated in a truly relational theory in any realm of science.

a. Relativity theory in physics still uses coordinate systems rather than the directly given spatial relations of entities. (The connection of relational methods in general to relativistic theories is not simple.)

b. As already mentioned (chapter 3, note 7) there is still no general theory of the role played by asymmetrical relations in the natural sciences or in any science, and hitherto the principle of increasing entropy could be combined with classical mechanics only by treating the *statistics* of assemblies, and the a priori *probabilities* of certain states as fundamental.

The consequence is that it is difficult to appreciate what a profound change in philosophy and science would result from a truly relational theory of knowledge in all or any realm. (Compare Russell, *Our Knowledge of the External World* [1914, 1922], pp. 45 51 and *passim.*) In my view it would mean that experienced qualities and measured quantities could both be derived from a single basis, particular qualities being experienced and quantities measured in certain contrasted circumstances. In fact, it appears, as suggested in the text, that changing patterns of spatial relations can cover both the (subjective) experience of qualities and the (objective) measurement of quantities. This agrees with I. Leclerc's view that "all characters, qualitative as well as quantitative, are characters of relation" (*The Nature of Physical Existence* [1972], p. 306). But these conjectures cannot be further developed here.

All that is relevant to this book is that a purely relational theory of knowledge would imply a basically changed metaphysics and ontology, that is, an altered view of the "real," of "existence," and so on. This general suggestion acquires precision when the relations involved are spatial, as here assumed. Nature is then the realm of changing forms, best defined by angles, and more often than not of morphic processes leading toward regular forms. Just as symmetrical relations played a primary role in the realm of quantity, so asymmetrical relations do in the realm of *order*.

In my view the mind/body problem will be transformed and partly resolved when primary emphasis is placed on what is common to mental and material systems: their hierarchical ordering (see chapter 3, note 6, and chapter 8). Important advances in intellectual understanding have often resulted from emphasis being placed on some highly specific factor, rather than on mere generalizations of earlier concepts. In my view the specific factor to which attention must now he paid, in order to clarify the mind/body problem, is the presence of relational, global, hierarchical, one way processes in both realms, requiring the use of asymmetrical relations.

It appears probable that the indispensable basic factor, underlying the current abstractions of advanced physical theory, which must eliminate the last vestiges of classical mechanism, is the presence of morphic processes in (3D, S), that is, in a succession of immediately observed patterns of 3D spatial relations. These require asymmetrical relations for their representation.

6. See L.L.W., *Critique of Physics* (1931), note 14, pp. 161-63.
7. The relativistic treatment of spatial and temporal relations as, in certain respects, equivalent and inseparable—if one neglects the necessity for an indefinite metric—is a consequence of the use of external coordinate systems, and is applicable only to relatively simple systems. In any treatment of complex systems using internal coordinates determined by the structure of the system, spatial relations are primary, and "time" enters only as a relation of temporal succession, for example:

$$At_2 < At_1, \text{ where } A = \text{some asymmetry}$$

(see chapter 3, note 7). This method implies, however, that instantaneous distances play a primary role. It seems that physics cannot do without a rigid rod, or the equivalent.
8. The derivation of the values of the relevant dimensionless constants as an integral part of the derivation of electron theory (e^2), quantum mechanics (h), and the Special Theory of Relativity (c), from a single principle. See *NDM*, chapter 11, p. 242.
9. For example see F. Manual, *Shapes of Philosophical History* (pp. 158—61). Astrologers, esoteric prophets, and occult visionaries have claimed to see a pattern in world history leading from our Dark Age to a Golden Age in the future. No rational evaluation of such ideas can be made today. In *NDM* I have expressed all that I can envisage of a possible future of mankind.

Chapter 5

1. In chapter 2, note 5, I considered points mainly of retrospective interest, dealing with the weaknesses of past theories of organism. Here I shall refer to recent publications concerned with the strengthening of biological theory.

Among these one of the most important, though no definitive clarifications are reached, is C. H. Waddington's *Towards n Theoretical Biology*, 4 vols. (1968 72). (The editor has kindly let me see parts of Volume IV in advance of publication.) I find the papers of Waddington, Wolpert, Puttee, and Lewontin of special interest. B. C. Goodwin's paper on "Biological Meaning" in Volume IV is also valuable. A reading of these four volumes gives one a picture of the condition of pioneer semiorthodox minds in biology today, but some important features of organisms are relatively neglected, for example, the chirality of biostructures, the role of internal factors in evolution, and the need for special organic variables. Wolpert has also written elsewhere on "Positional Information and Pattern Information" in *Current Topics in Developmental Biology*, vol. 6 (1971).

A development which I consider likely to provide insights of importance is the work of D. H. Habel and T. N. Wiesel (of Harvard), *Journal of Physiology*, 118 [1959], 574 91; 160 [1962], 106, and later. This and parallel investigations may be leading toward the identification of a hierarchical ordering of the detector cells in the striate cortex, which could conceivably result, perhaps indirectly, in a correlation of hierarchical forms in brain physiology with those in cognitive processes (see chapter 8).

When writing *Internal Factors in Evolution* (1965) I had the satisfaction of discovering that I was one of seven thinkers who had called attention to the importance of internal factors during the four years 1949–53, as this supported my views and proved that they were timely. It was also interesting to learn that Charles Darwin in a letter (1858) referred to the analysis of the status of structures neither beneficial nor injurious as "one of the greatest oversights yet detected in my work." I owe this reference to Loren Eiseley.

If there were more who combined interests in physical theory, molecular biology, and in evolutionary theory, I would expect increased attention to be paid to the role of chiral structures in organic coordination, and therefore also in biopoesis.

In ethology, a development has attracted my interest which stresses the role of attention (a feature at once physiological and cognitive) in animal behavior. Recent workers here placed the main emphasis in studying the group behavior of higher primates on the *communication of signals* within a group. M. R. A. Chance (*Man*, II, New Series, 4 [1967], 503; and subsequent publications) has raised a prior issue: *How is a group maintained?* He suggests the answer: In most cases *by a pattern of attention within the group*. A pattern of attention focused on one individual, for example, by all members of a group continually watching a dominant male, alone makes possible a pattern of group intercommunication. The coherence of the group is maintained by almost continuous attention during daylight to the behavior of one dominant member. This interpretation has been supported by observations made on several primates.

I wish also to call attention to the work of W. H. Thorpe and his colleague R. A. Hinde on bird vocalization and animal behavior, though here I am relying on the judgment of others. Convenient references are R. A. Hinde, ed., *Bird Vocalisations* (1969), and R. A. Hinds, ed., *Non-verbal Communication* (1972). But as far as I know there does not appear to be yet any adequate interpretation of the role of heredity in relation to animal behavior, and this may not be possible until the morphogenesis of the central nervous system is better understood. A good summary of the situation is R. A. Hinde's *Animal Behavior* (1970).

2. For example, in superconductivity, in lasers, and in the Mossbauer effect. It is noteworthy that early quantum mechanics failed to pay adequate attention to the properties of highly ordered systems. As far as I know no one noticed at the time.
3. On "coordinative conditions" (C.C.) see L.L.W., *Internal Factors in Evolution*, chapter 1. Chiral properties probably play an important role in the C.C., since emantiomers are rarely found mixed in cellular organisms. This suggests that the presence of structures of one chiral sense only in any situation is, at least at the molecular level, an indispensable condition of organic stability and coordination. Living controls, it seems, require chirality of one sense only in any one region at any one level. "Life is a linked set of reactions, and therefore their component molecules must depend on fitting their chirality, right. and left handedness together." J. D. Bernal, *The Origin of Life* (1967) , p. 144.
4. More precisely, *nonholonomic constraints*, which reduce the effective number of degrees of freedom as compared with the number assumed to be present in the representation used. Thus the number of constraints present depends on the representation selected. H. Hertz introduced this conception in his *Principles of Mechanics* (1894). It is possible that the representation of molecular systems which possess, within certain energy thresholds, a greatly reduced number of degrees of freedom owing to chemical bonds, etc., at various levels, may require an extended or strengthened form of quantum mechanics, since deformable molecules subject to such constraints may not be covered by its standard methods. See H. H. Pattee, "The Pbysical Basis of Coding," in Vol. I of Waddington's *Towards a Theoretical Biology* (1968).
5. See Pattee, preceding note.
6. See Wolpert, note 1 above.
7. On pulsations, see L.L.W., "Structural Philosophy of Organism," *B.J. Phil. Sci.*, V (1955), 332. A pulsation involves the equivalent of a thermodynamic cycle in a single complex structure. See discussion of McClare, ehapter 4, note 4b, above. Also A. S. Eddington, *The Internal Constitution of the Stars* (1926), p. 56, where he

stresses, in a treatment of the quantum theory of the interactions of radiation and matter, that "formally at least we are balancing cycles of processes instead of a direct and reversed process." More than formally, I suggest: there is *never* detailed balancing of direct and inverse processes; they occur in different systems, involve different up and down paths and a thermodynamic cycle, or the equivalent. It is clear that they must occur in different systems, and therefore involve distinct paths, since when receiving energy the structure must be in interaction with an energy source, and when relaxing in interaction with some different structure on which it does work.

Chapter 6

I recommend the following somewhat arbitrary selection of books throwing light on the present situation of mankind, or of the United States as a symptom of it. They are of varied quality, but each of them, in my view, contributes something important. However, they all appear to me to neglect the role of the unconscious tradition in developing a possible consensus tomorrow, and none of them has led me to modify my emphasis on the world view.

F. C. S. Northrop, *The Meeting of East and West* (1946)
W. W. Wagar, *The City of Man* (1963)
W. W. Wagar, *Building the City of Man* (1971)
J. R. Platt, *The Step to Man* (1966)
T. Roszak, *The Making of a Counter Culture* (1969)
C. A. Reich, *The Greening of America* (1970)
to which I add:
L.L.W., *The Next Development in Man* (1944)

One should not neglect past authors who have in some degree foreseen the present situation. Their views can improve our perspective, and their insights may in some respects be better than our own, since we are confused by the complexity and intensity of current problems. On P. B. Shelley's insights, see the last chapters of C. Small, *Ariel Like a Harpy* (1972). Anyone who digests and understands these eight books will, in my view, be assisted in interpreting the present situation.

1. Thomas Hardy, *Jude the Obscure* (1895).
2. M. Polanyi, *Beyond Nihilism*, Eddington Lecture (1960).
3. At least two thinkers have suggested that the experience of free will involves a mental process at one level, and the recognition of necessity a process at another level. Thus "*the shift of control* of an outgoing activity from one level to a higher level of the hierarchy—from 'mechanical' to 'mindful' behavior—seems to be the essence of conscious decision making and of the subjective experience of free will." A. Koestler in *Beyond Reductionism* (1969), p. 206; see also *The Ghost in the Machine*, p. 208. A similar idea is expressed in I. A. Richards's *Coleridge on the Imagination* (1934), p. 62. This interpretation in terms of levels is clearer than F. Selman's vague assertion that predestination and free will express "opposite sides of a single reality."
4. This is one of Jean Piaget's central ideas. See his *Biology and Knowledge* (1971), p. 26 and *passim*, and other works. Others have assumed this, but Piaget gives it a special emphasis in his genetic epistemology. Piaget's work provides support for the themes of chapters 5 and 6.

5. L.L.W., "Goethe's Single View of Nature and Man," *German Life and Letters*, 2 (July 1949), 287. Also in *NDM*.
6. L.L.W., "Towards a Science of Form," dedicated to Herbert Read, *Hudson Review*, XXIII (Winter 1970 71), 613.
7. I cannot give references to those who have interpreted "mystical experience" in a naturalistic manner. M. Laski has given a useful approach in her *Ecstasy*, 1961, pp. 369–74.
8. I have not space here to develop in detail my view that in any theory of mental capacities, the imagination, properly defined, must play a primary role, and most other capacities be derivable as special cases of, or aspects of, imaginative activity. The imagination, since it develops the new, requires the relation of succession and therefore has to be represented by asymmetrical relations; and the analytical intellect, since it normally deals with invariant *terms* of symmetrical relation, must, in my view, be derivable from the former.

   I regard the imagination as a perconscious process forming unified orderings of relations, often but not always of a kind meeting standard needs of the organism, and the analytical intellect as normally concerned with (nearly) invariant terms of such relations. The imagination can be defined and understood without the intellect; but any understanding of the intellect must assume some ordering generated by the imagination.
9. I am probably prejudiced in favor of the U.S.A. I am in a sense part American, for my paternal grandfather fought at Bull Run, having answered Lincoln's first call for volunteers. Several of my brothers and sisters have known North America well. My moral and financial support over the last twenty years have come to me mainly from the United States, and I seize this chance of saying, "Thank you, young in spirit U.S.A., foundations and individuals."

   But putting this bias apart as best I can, I consider that the future of mankind in this century will be more influenced by the young of the U.S.A. than by any other comparable group. More than that, if mankind comes through to hail the year 2000 *it will owe this primarily to a new U.S.A.*, now coming into being behind all the ugliness, violence, and dominance of money, and since 1950 putting a fresh emphasis on the quality of individual experience.

   I could develop this theme further. Instead I ask those who wish to understand this somewhat paradoxical hope to read Harrison Brown's "An American Renaissance" (see note 10 below). At least ten or twenty years must be allowed to pass before this evaluation of the role of young inspirit America can be seen to be correct or not.

   The term "revolution" may have misleading implications, but I agree with J. F. Revel, in his *Ni Marx ni Jesus* (Paris, 1970), that a world "revolution" has already begun (that is, a metamorphosis of the psyche), with the U.S.A. as its dominant focus, which will ultimately have international, political, social, technological, and cultural consequences that we cannot yet foresee. This is in line with *NDM* (p. 256), where I stated that the most important historical instrument of the global development will be the U.S.A., but also gave a crucial role to pressure from Asia.
10. In my view ethical philosophers and philosophers of history have been too little interested in changes of opinion and so have until recently underestimated the important, indeed sometimes decisive, role of the press in recent times and even of the new communications media, obvious though that is now. I have often noted that a very few writers, say five to ten, have prepared the way for a necessary and ultimately decisive switch of public opinion in England or America. These few are the salt of the earth, but they are usually given their opportunity by editors or owners of the media. At the present moment the existence in the Western world of some hundred responsible men of good will and good judgment—if they exist in posi-

tions of influence today—facilitating the spread of a world consensus by selecting appropriate reporters and commentators could be of decisive influence. Subject to the theme of chapter 7, which lies at a deeper and ultimately overriding level, the following appears to me to be probably true: a few hundred newswriters and commentators now have in their hands human fate in this century. Great themes must be brought to birth by a few. But chapter 7 throws a deeper light on the role of these few.

This is so important that I will give two examples. The recognition of Hitler as antihuman in the deepest sense by England and the United States was facilitated just in time by a mere handful of newspapermen, mainly American. This is a terrifying thought. Perhaps even more startling is the fact that one of the most valuable journals in the West, *The Bulletin of Atomic Scientists* (Chicago), is editorially said to be in danger of ceasing publication through lack of finance. This periodical has for a generation been seeking to identify sane social policies in a time dominated by technology and the bomb. If justice were relevant, one would say that a civilization which could not keep such a journal alive deserves to go down. The U.S.A. is choked with apparently well meaning foundations. Why are they so blind?

This bulletin is now called *Science and Public Affairs*, and its format and contents have been slightly changed. But its aim and quality remain the same. My views about the Bulletin are so strong that my expression of them may seem to be exaggerated. I know of no other journal in the world, certainly not in the English speaking realm, which has consistently over so long done so much toward the enlightenment of mankind on what many consider to be the supreme short term issue: *science, ethics, and mankind.* I will be more specific. The June 1972 issue contained one brilliant paper by Harrison Brown, on "An American Renaissance." This is a noble voice speaking from the most powerful and vital country in the world. Some foundation should send a copy of this article to every American college student and every administrator under forty years, so as not to waste paper on the already over-successful, well satisfied, and largely indifferent oldsters. (I confess I am tired of making such suggestions to the deaf. Perhaps Young America can do something about it.) If Harrison Brown's suggestions became not only official American policy, but were actually implemented, there is a good chance that mankind would be saved. Nearly every issue of the *Bulletin* has contained at least one paper of similar quality, though on narrower themes.

The *Bulletin* was created by the United States physicists' sense of responsibility for having made the bomb. I have asked the following question several times over 25 years, but it was apparently too uncomfortably penetrating and I received no answers to several letters: *Have West German biologists created a comparable journal seeking to show mankind how to avoid a repetition o/ the Hitler genocide, which was supported by some of them?* Why not? Are their consciences clear? Life is too brief for me to forgive a Germany which has not even tried to do what American scientists are struggling to achieve. German biology carries a heavy load of guilt, for which they have not yet atoned. But now the responsible individuals are dying off, and it will soon be too late.

Chapter 7

1. See L.L.W., *The Unconscious before Freud* (1960).
2. See chapter 4, note 4c.
3. Goethe translation: "To the Moon": What is unknown and unthought by man wanders in the night through the labyrinth of the heart.

4. I must condense into this note the evidence that mankind is now changing its mind rather suddenly, at least in some areas.

a. In a period of fifteen years, 1955–70, the *previously dominant ethos* in *the U.S.A.* (including its conception of its own role in the world) *collapsed with unprecedented suddenness*. A new generation of writers became influential, many youth and colored protest movements became active, several books expressed a new attitude, and American foreign policy was compelled to begin the retreat from Vietnam. So sudden a collapse of ethos and reversal of policy may never before have occurred in a major world power. What was revolutionary and un-American in 1940–55 became a commonplace by 1970: *the United States knows it is misbehaving* in many respects. Two books already referred to above, whatever their limitations (I do not entirely agree with either), expressed the new awareness unmistakably: T. Roszak's *Making of a Counter Culture* and C. Reich's *Greening of America*. It would be false modesty not to point out that my *NDM* (1944) foretold some of the features described by these two authors, but it stressed the collapse of a dissociated and the emergence of a new unified awareness, as an idea to be lived and worked for, without seeking to estimate the probabilities of success.

b. The suddenly increased attention to *pollution* is another example.

c. The recent rapid growth of interest in the *occult* astonishes and alarms some. But thus is an inevitable reaction from the false doctrine of mechanism and the dogmatic materialism and utilitarianism of recent science and of social policy. The universe does not obey the mechanical laws assumed up to around 1870, but some scientists and philosophers still treat "mechanism" as if it were a fundamental concept or principle of exact science. Who knows what surprising truth may not lie in the realm now called "occult"? Here I am personally cautious and skeptical, but again I must expect surprises precisely where they seem to me to be improbable.

The near simultaneity of these three sudden changes is not extraordinary, since they all represent aspects of one major change in the Western psyche, and possibly of mankind.

As suggested in the text the outburst of a new idea or attitude is usually marked by: (a) suddenness due to the swift passing of a threshold; (b) simultaneity between different local expressions, inexplicable without the assumption of an unconscious tradition or the equivalent; and (c) nearly simultaneous outbursts concerning various particular attitudes, which are actually the expression of a single underlying metamorphosis. In spite of the widespread evidence of a change, there is often one dominant focus in which the metamorphosis is expressed early and most intensely.

5. The most important of these were, approximately in chronological sequence: *de Broglie, Born,* Bose, *Einstein, Schrod¿nger, Heisenberg,* Fermi, *Dirac, Pauli,* Jordan, Wiener, *Wigner,* Klein, and Gordon. (I have italicized eight that I consider of the first rank.) During the seven years 1923–29 these 14 made important contributions to fundamental quantum mechanics which were coordinated into a unified theory only around 1930 by Dirac and others. Some of these men worked in pairs, but at least eight seemingly entirely independent contributions were made whose relationships as components of one highly integrated theory were understood only at the end, around 1929–30.1 know of no adequate discussion of this extraordinary occurrence, which is unique in the history of human thought. Perhaps one of those few who are still alive will put on record his retrospective interpretation of how his particular contribution came to be coherent with those of so many others.

6. See R. Lannoy, The Speaking Tree: *A Study of the Indian Culture and Society* (Oxford, 1971), pp. 270 74, 283, 384 85, 412 ff. My interpretation of his book and his references to my writings support the theme of the text.
7. E. v. Hartmann, *Philosophy of the Unconscious* (1868), chapter 10.
8. (L. L. Whyte had intended to discuss in more detail elsewhere "the now indispensable, potentially universal consensus of heart, mind, and will which all nonpathological individuals must necessarily accept, that alone can lead mankind safely through the next few decades. I propose there to treat the role of the unconscious tradition in carrying this consensus over the threshold from unconscious levels into the fully conscious mind, the signals of this developing consensus that may be expected, the phases to be passed through, the role of individuals and communities, the means by which the emergence of the consensus can best be facilitated, and the way in which local diversity can be maintained in spite of this overriding drive toward a unification sufficient to save the species. As has already been suggested by some, an overriding unity can sometimes facilitate the maximum diversity.")

Chapter 8

1. I have neither experienced nor read much in this realm, but I recommend Rosalind Heywood's two paperbacks: *The Sixth Sense: A General Enquiry into ESP* (1959) and *The Infinite Hive: A Personal Record of Extra. Sensory Experiences* (1964). I know Mrs. Heywood to be a person whose statements about facts can be trusted. I also recommend Arthur Koestler's *Roots of Coincidence* (1972) as a lively introduction to ESP and its present status. [A. Hardy, R. Harvie, and A. Koestler, *The Challenge of Chance: Experiment and Speculations* (London: Hutchinson, 1973) was published since this was written.]

Index